PROGRAMMING

Seth Lloyd is Professor of Quantum-Mechanical Engineering at the Massachusetts Institute of Technology. His research focuses on complexity and on how physical systems process information. He was the first person to develop a realisable model for quantum computation and is currently working with scientists around the world to construct and operate quantum computers and quantum communication systems.

5 | 2069138

SETH LLOYD

Programming the Universe

A Quantum Computer Scientist
Takes on the Cosmos

VINTAGE BOOKS
London

Published by Vintage 2007

2 4 6 8 10 9 7 5 3 1

Copyright © Seth Lloyd 2005

Seth Lloyd has asserted his right under the Copyright, Designs
and Patents Act, 1988 to be identified as the author of
this work

This book is sold subject to the condition that it shall not
by way of trade or otherwise, be lent, resold, hired out, or
otherwise circulated without the publisher's prior consent in any
form of binding or cover other than that in which it is published
and without a similar condition including this condition being
imposed on the subsequent purchaser

First published in Great Britain in 2006 by
Jonathan Cape
Random House, 20 Vauxhall Bridge Road,
London SW1V 2SA

www.vintage-books.co.uk

Addresses for companies within The Random House Group Limited
can be found at: www.randomhouse.co.uk/offices.htm

The Random House Group Limited Reg. No. 954009

A CIP catalogue record for this book
is available from the British Library

ISBN 9780099455370

The Random House Group Limited makes every effort to
ensure that the papers used in its books are made from trees that
have been legally sourced from well-managed and credibly certified
forests. Our paper procurement policy can be found at:
www.randomhouse.co.uk/paper.htm

Printed and bound in Great Britain by
Cox & Wyman Limited, Reading, Berkshire

To Eve

Contents

The Apple and the Universe

"In the beginning was the bit," I began. The chapel in the seventeenth-century convent that housed the Santa Fe Institute for the study of complex systems was filled with the usual collection of physicists, biologists, economists, and mathematicians, with a leavening of Nobel laureates. The grand old man of astrophysics and quantum gravity, John Archibald Wheeler, had challenged me to give a talk on the subject "It from Bit." I had accepted the challenge. I was beginning to wonder if that had been a good idea, but it was too late to back down now. I held an apple in my hand.

"Things, or 'its,' arise out of information, or 'bits,' " I continued, nervously tossing the apple in the air. "This apple is a good 'it.' Apples have long been associated with information. First of all, the apple is the fruit of knowledge 'whose mortal taste brought death into the world, and all our woe.' It conveys information about good and evil. Down the line, it was in the trajectory of a falling apple that Newton traced the universal laws of gravitation, and the curved surface of the apple is a metaphor for Einstein's curved spacetime. More directly, the genetic code locked in the seeds of an apple programs the structure of future apple trees. And last, but not least, an apple contains free energy—the calories of bit-rich energy that our bodies need to function." I took a bite of the apple.

"Clearly, there are many *types* of information contained in this apple. But *how much* information does the apple embody? *How many* bits are there in an apple?" I placed the apple on the table and turned to the board to perform a short calculation. "Interestingly, the number of bits in an apple has been known since the beginning of the twentieth century, since before the word 'bit.' At first, one might think that an apple embodies an infinite number of bits, but this is not so. In fact, the laws of quantum mechanics, which govern all physical systems, make finite the number of bits required to specify the microscopic state of the apple and its atoms. Each atom, by its position and velocity, registers only a few bits; each nuclear spin in an atom's core registers but a single bit. As a result, the apple contains only a few times more bits than atoms—a few million billion billion zeros and ones."

I turned back to face the audience. The apple was gone. Not good. Who had taken it? I glanced at the benign face of Wheeler and the impassive expression of Murray Gell-Mann, Nobel laureate, inventor of the quark and wearer of one of the world's physics heavyweight title belts.

"I can't continue without the apple. No it, no bit," I declared, and sat down.

The hunger strike lasted only a few moments before an impish engineer from Bell Labs produced the apple. I took it from him and held it on high, issuing a challenge to anyone who might attempt another theft. This was a mistake. For the moment, though, all seemed well. I continued:

"All bits are equal in terms of the amount of information they can register. A bit, short for 'binary digit,' is registered by two distinguishable states—0 or 1, yes or no, heads or tails. Any physical system with two such states registers exactly one bit. A system with more states registers more bits. A system with four states—for example, 00, 01, 10, 11—registers two bits; a system with eight states—for example, 000, 001, 010, 011, 100, 101, 110, 111—registers three bits, and so on. As mentioned before, quantum mechanics guarantees that any physical system with finite energy confined to a finite volume of space has a finite number of distinguishable states and therefore will register a finite number of bits. All physical systems register information. In the words of IBM's Rolf Landauer, 'Information is physical.' "

Here Gell-Mann interrupted: "But are all bits truly equal? What about the bit that tells us whether some famous unsolved mathematical conjecture is true or not? Compare that with a bit derived from a random coin toss. Some bits are more important than others."

True, I agreed. Different bits play different roles in the universe. All bits can register the same amount of information, but the quality and importance of that information varies from bit to bit. The significance of "yes" depends on the question asked. The two bits of information that determine the identity of a base pair in the apple's DNA are far more important for generations of future apples than the bits of information registered by the thermal jiggling of a carbon atom in one of the apple's molecules. Only a few molecules and their attendant bits are required to convey the smell of the apple, whereas billions of billions of bits are needed to provide the apple with its nutritional value.

"But," Gell-Mann interjected, "is there a mathematically precise way of quantifying the significance of a bit?"

I did not have a complete answer to this question, I replied, still holding the apple. The significance of a bit of information depends on how that information is processed. All physical systems register information, and when they evolve dynamically in time, they transform and process that information. If an electron "here" registers a 0 and an electron "there" registers a 1, then when the electron goes from here to there, it flips its bit. The natural dynamics of a physical system can be thought of as a computation in which a bit not only registers a 0 or 1 but acts as an instruction: 0 can mean "do this" and 1 can mean "do that." The significance of a bit depends not just on its value but on how that value affects other bits over time, as part of the continued information processing that makes up the dynamical evolution of the universe.

I continued to identify the bits from which the apple arises and to elaborate the roles those bits play in the processes that make up the apple's characteristics. Things were going well. I had addressed the problem of "it from bit" and had survived the questioning. Or so I thought.

As I finished the talk and stepped away from the board, someone tackled me from behind. One of the audience members had taken seriously my challenge to steal the apple. Doyne Farmer was one of the founders of chaos theory—a tall, athletic man. He grabbed my arms to make me drop the apple. To break his grasp, I slammed him back against the wall. Pictures of fractals and photos of pueblos fell. But before I could wriggle free, Farmer wrestled me to the ground. We rolled around the floor, overturning chairs. By now, the apple was gone. It had been reduced to bits.

Part 1

THE BIG PICTURE

CHAPTER 1

Introduction

This book is the story of the universe and the bit. The universe is the biggest thing there is and the bit is the smallest possible chunk of information. The universe is made of bits. Every molecule, atom, and elementary particle registers bits of information. Every interaction between those pieces of the universe processes that information by altering those bits. That is, the universe computes, and because the universe is governed by the laws of quantum mechanics, it computes in an intrinsically quantum-mechanical fashion; its bits are quantum bits. The history of the universe is, in effect, a huge and ongoing quantum computation. The universe is a quantum computer.

What does the universe compute? It computes itself. The universe computes its own behavior. As soon as the universe began, it began computing. At first, the patterns it produced were simple, comprising elementary particles and establishing the fundamental laws of physics. In time, as it processed more and more information, the universe spun out ever more intricate and complex patterns, including galaxies, stars, and planets. Life, language, human beings, society, culture—all owe their existence to the intrinsic ability of matter and energy to process information. The computational capability of the universe explains one of the great mysteries of nature: how complex systems such as living creatures can arise from fundamentally simple physical laws. These laws allow us to predict the future, but only as a matter of probability, and only on a large scale. The quantum-computational nature of the universe dictates that the details of the future are intrinsically unpredictable. They can be computed only by a computer the size of the universe itself. Otherwise, the only way to discover the future is to wait and see what happens.

Allow me to introduce myself. The first thing I remember is living in a chicken house. My father was apprenticed to a furniture maker in Lincoln, Massachusetts, and the chicken house was in back of her barn. My father turned the place into a two-room apartment; the space where the chickens had roosted became bunks for my older brother and me. (My younger brother was allowed a cradle.) At night, my mother would sing to us, tuck us in, and close the wooden doors to the roosts, leaving us to lie snug and stare out the windows at the world outside.

My first memory is of seeing a fire leap up in a wire trash basket with an overlapping diamond pattern. Then I remember holding tight to my mother's blue-jeaned leg just above the knee and my father flying a Japanese fighter kite. After that, memories crowd on thick and fast. Each living being's perception of the world is unique and crowded with detail and structure. Yet we all inhabit the same space and are governed by the same physical laws. In school, I learned that the physical laws governing the universe are surprisingly simple. How could it be, I wondered, that the intricacy and complexity I saw outside my bedroom window was the result of these simple physical laws? I decided to study this question and spent years learning about the laws of nature.

Heinz Pagels, who died tragically in a mountaineering accident in Colorado in the summer of 1988, was a brilliant and unconventional thinker who believed in transgressing the conventional boundaries of science. He encouraged me to develop physically precise techniques for characterizing and measuring complexity. Later, under the guidance of Murray Gell-Mann at Caltech, I learned how the laws of quantum mechanics and elementary-particle physics effectively "program" the universe, planting the seeds of complexity.

These days, I am a professor of mechanical engineering at the Massachusetts Institute of Technology. Or, because I have no formal training in mechanical engineering, it might be more accurate to call me a professor of quantum-mechanical engineering. Quantum mechanics is the branch of physics that deals with matter and energy at its smallest scales. Quantum mechanics is to atoms what classical mechanics is to engines. In essence: I engineer atoms.

In 1993, I discovered a way to build a quantum computer. Quantum computers are devices that harness the information-processing ability of individual atoms, photons, and other elementary particles. They compute in ways that classical computers, such as a Macintosh or a PC, cannot. In the process of learning how to make atoms and molecules—the smallest pieces of the universe—compute, I grew to appreciate the intrinsic information-processing ability of the universe as a whole. The complex world we see around us is the manifestation of the universe's underlying quantum computation.

The digital revolution under way today is merely the latest in a long line of information-processing revolutions stretching back through the development of language, the evolution of sex, and the creation of life, to the beginning of the universe itself. Each revolution has laid the groundwork for the next, and all information-processing revolutions since the Big Bang stem from the intrinsic information-processing ability of the universe. The computational universe *necessarily* generates complexity. Life, sex, the brain, and human civilization did not come about by mere accident.

The Quantum Computer

Quantum mechanics is famously weird. Waves act like particles, and particles act like waves. Things can be in two places at once. It is perhaps not surprising that, at small scales, things behave in strange and counterintuitive ways; after all, our intuitions have developed for dealing with objects much larger than individual atoms. Quantum weirdness is still disconcerting, though. Niels Bohr, the father of quantum mechanics, once said that anyone who thinks he can contemplate quantum mechanics without getting dizzy hasn't properly understood it.

Quantum computers exploit "quantum weirdness" to perform tasks too complex for classical computers. Because a quantum bit, or "qubit," can register both 0 and 1 *at the same time* (a classical bit can register only one or the other), a quantum computer can perform millions of computations simultaneously.

Quantum computers process the information stored on individual atoms, electrons, and photons. A quantum computer is a democracy of information: every atom, electron, and photon participates equally in registering and processing information. And this fundamental democracy of information is not confined to quantum computers. All physical systems are at bottom quantum-mechanical, and all physical systems register and process information. The world is composed of elementary particles—electrons, photons, quarks—and each elementary piece of a physical system registers a chunk of information: one particle, one bit. When these pieces interact, they transform and process that information, bit by bit. Each collision between elementary particles acts as a simple logical operation, or "op."

To understand any physical system in terms of its bits, we need to understand in detail the mechanism by which each and every piece of that system registers and processes information. If we can understand how a quantum computer does this, then we can understand how a physical system does.

The idea of such a computer was proposed in the early 1980s by Paul Benioff, Richard Feynman, David Deutsch, and others. When they were first discussed, quantum computers were a wholly abstract concept: Nobody had a clue how to build them. In the early 1990s, I showed how they could be built using existing experimental techniques. Over the past ten years, I have worked with some of the world's greatest scientists and engineers to design, build, and operate quantum computers.

There are a number of good reasons to build quantum computers. The first is that we can. Quantum technologies—technologies for manipulating matter at the atomic scale—have undergone remarkable advances in recent years. We now possess lasers stable enough, fabrication techniques accurate enough, and electronics fast enough to perform computation at the atomic scale.

The second reason is that we have to—at least if we want to keep building ever faster and more powerful computers. Over the past half century, the power of computers has doubled every year and a half. This explosion of computer power is known as "Moore's law," after Gordon Moore, subsequently the chief executive of Intel, who noted its exponential advance in the 1960s. Moore's law is a law not of nature, but of human ingenuity. Computers have gotten two times faster every eighteen months because every eighteen months engineers have figured out how to halve the size of the wires and logic gates from which they are constructed. Every time the size of the basic components of a computer goes down by a factor of two, twice as many of them will fit on the same size chip. The resulting computer is twice as powerful as its predecessor of a year and half earlier.

If you project Moore's law into the future, you find that the size of the wires and logic gates from which computers are constructed should reach the atomic scale in about forty years; thus, if Moore's law is to be sustained, we must learn to build computers that operate at the quantum scale. Quantum computers represent the ultimate level of miniaturization.

The quantum computers my colleagues and I have constructed already attain this goal: each atom registers a bit. But the quantum computers we can build today are small, not only in size but also in power. The largest general-purpose quantum computers available at the time of this writing have seven to ten quantum bits and can perform thousands of quantum logic operations per second. (By contrast, a conventional desktop computer can register trillions of bits and can perform billions of conventional, classical logic operations per second.) We're already good at making computers with atomic-scale components; we're just not good at making *big* computers with atomic-scale components. Since the first quantum computers were constructed a decade ago, however, the number of bits they register has doubled almost every two years. Even if this exponential rate of progress can be sustained, it will still take forty years before quantum computers can match the number of bits registered by today's classical computers. Quantum computers are a long way from the desktop.

The third reason to build quantum computers is that they allow us to understand the way in which the universe registers and processes information. One of the best ways to understand a law of nature is to build and operate a machine that illustrates that law. Often, we build the machine first and the law comes later. The wheel and the top had existed for millennia before the establishment of the law of conservation of angular momentum. The thrown rock preceded Galileo's laws of motion; the prism and the telescope came before Newton's optics; the steam engine preceded James Watt's governor and Sadi Carnot's second law of thermodynamics. Since quantum mechanics is so hard to grasp, wouldn't it be nice to build a machine that embodies the laws of quantum mechanics? By playing with that machine, one could acquire a working understanding of quantum mechanics, just as a baby who plays with a top grasps the principles of angular momentum embodied by the toy. Without direct experience of how atoms actually behave, our understanding remains shallow. The "toy" quantum computers we build today are machines that will allow us to learn more and more about how physical systems register and process information at the quantum-mechanical level.

The final reason to build quantum computers is that it's fun. In the pages to come, you'll meet some of the world's foremost scientists and engineers: Jeff Kimble of Caltech, constructor of the world's first photonic quantum logic gate; Dave Wineland of the National Institute of Standards and Technology, who built the first simple quantum computer; Hans Mooij of the Delft University of Technology, whose group gave some of the earliest demonstrations of quantum bits in superconducting circuits; David Cory of MIT, who built the first molecular quantum computer, and whose quantum analog computers can perform computations that would require a classical computer larger than the universe itself. Once we have seen how quantum computers work, we will be able to put bounds on the computational capacity of the universe.

The Language of Nature

As it computes, the universe effortlessly spins out intricate and complex structures. To understand how the universe computes—and thus to understand better those complex structures—we must learn how it registers and processes information. That is, we must learn the underlying language of nature.

Think of me as a kind of atomic masseur. As a professor of quantum-mechanical engineering at MIT, my job is to massage electrons, photons, atoms, and molecules into those special states in which they become quantum computers and quantum communication systems. Atoms are tiny but strong, resilient but sensitive. They are easy to talk to (just hit the table and you've talked to billions upon billions of them) but hard to listen to (I bet you can't tell me what the table had to say beyond "thump"). They don't care about you, and they go about their business doing what they have always done. But if you massage them in just the right way, you can charm them. They will compute for you.

Atoms are not alone in their ability to process information. Photons (particles of light), phonons (particles of sound), quantum dots (artificial atoms), superconducting circuits—all these microscopic systems can register information. And if you speak their language and ask them nicely, they will process that information for you. What language do such systems speak? Like all physical systems, they respond to energy, force, and momentum, to light and sound, to electricity and gravity. *Physical systems speak a language whose grammar consists of the laws of physics.* Over the last ten years, we have learned this language well enough to talk to atoms—to convince them to perform computations and report the results.

How hard is it to "speak Atom"? To learn to converse fluently takes a lifetime. I myself am a poor atomic conversationalist, compared with other scientists and quantum-mechanical engineers you will meet in this book. To learn enough to carry on a simple conversation, however, is not hard.

Like all languages, Atom is easier to learn when you're younger. With Paul Penfield, I co-teach a freshman course at MIT called Information and Entropy. The goal of this course, like the goal of this book, is to reveal the fundamental role that information plays in the universe. Fifty years ago, MIT freshmen used to arrive full of knowledge about internal-combustion engines, gears and levers, drivetrains and pulleys. Twenty-five years ago, they arrived full of knowledge of vacuum tubes, transistors, ham radios, and electronic circuits. Now they arrive chock-a-block full of knowledge about computers, disk drives, fiber optics, bandwidth, and music- and image-compression codes. Their predecessors lived in worlds dominated by mechanical and electrical technologies; today's freshmen come from a world dominated by information. Their predecessors already knew lots about force and energy, voltage and charge; today's freshmen already know lots about bits and bytes. The freshmen in our course already know so much about information technology that we can teach them subjects—including quantum computation—that previously could be taught only to graduate students. (My senior colleagues in the Mechanical Engineering Department complain that the incoming freshmen have never used a screwdriver. This is untrue. Fully half of them have used a screwdriver to install more memory in their computers.)

As part of a research project supported by the National Science Foundation, I have developed a class to teach first- and second-graders about how information is processed at the microscopic scale. Six- and seven-year-olds these days also come equipped with scarily extensive knowledge of computers. They, too, seem to have no trouble learning about bits and bytes. When asked to play a game in which each student represents an atom in a quantum computer, they do so readily and accurately.

Those of us who grew up before the current information-processing revolution, though, have no less appreciation for the variety and significance of information than do our bit-saturated juniors. Old or young, by the time you finish reading this book, you will know how to ask atoms to perform simple computations, employing machines that are available throughout the world, along with the grammar of the language of nature.

Information-Processing Revolutions

The underlying ability of the universe to process information has given rise, through history, to a series of information-processing revolutions. We are in the midst of such a revolution now, one driven by the rapid advance of the electronic computing technology embodied by Moore's law. Quantum computers represent the avant-garde of this revolution. But exciting and tumultuous as it is, ours is neither the first nor the greatest revolution in information processing.

Just as great a revolution was the invention of zero. The number zero was invented by the ancient Babylonians and passed down through the Arab world. The use of zero to represent powers of ten (10, 100, 1,000, etc.) distinguishes our Arabic number system from number systems such as that of the Romans, which used distinct symbols for powers of 10 (X = 10; C = 100; M = 1,000). Though it might seem only a slight change in the form of numerical representation, the invention of Arabic numbers has had a significant impact on mathematical information processing. (Not the least of which was the improvement in ease and transparency of commercial transactions. If the folks at Enron had been able to do their shady accounting in Roman numerals, they might have gotten away with it!) The origins of the Arabic number system are indistinguishable from the origins of its accompanying technology, the abacus—a simple, robust, and powerful calculating machine that consists of rows of movable beads mounted on sticks. The first row corresponds to ones, the second to tens, the third to hundreds, and so on. An abacus with just ten rows can perform calculations ranging into the billions.

Even more powerful than the ability of the abacus to deal easily with

Figure 1a. The Ascent of Bits

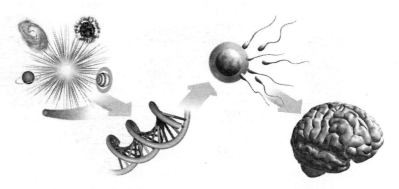

*The history of the universe can be thought of as a sequence of informa-
tion processing revolutions, each of which builds on the technology of
the previous ones.*

large numbers is its embodiment of the concept of zero. In fact, it seems
likely that the machine predated the word. The word "zero" is Italian,
short for *zefiro*, from the Low Latin *zephirum*, Old French *cifre*, Arabic *sifr*,
Sanskrit *shunya*—"an empty thing." In the Arabic number system, zero
acts as a placeholder, allowing ever larger numbers (10, 100, 1,000, . . .) to
be expressed with ease. An empty thing is a powerful device. In spite of—
or possibly because of—its power, zero as a number arouses suspicion. It
is somehow unnatural. Indeed, it is not one of the natural numbers (1, 2,
3, . . .). Zero in the abstract is a thorny concept, but an abacus with all
beads down is a simple, concrete thing: zero.

The abacus shows how a revolution in information processing cannot
be separated from the underlying mechanism or technology that governs
how the information is represented and processed. The information-
processing technology (e.g., the abacus) is typically inseparable from the
conceptual breakthrough (e.g., zero).

Going thousands of years farther back, we find an even greater revolution: writing. The original technology consisted of scratching marks

Figure 1b. The Ascent of Bits

on clay or rock. Writing was, almost literally, language made concrete. It enabled large-scale social organization, contracts, scriptures, and books like this one. Over the years, the technologies of writing have progressed from rock to paper to electrons. Each manifestation of writing, from commandment to poem to neon sign, possesses its own variation on the technology for representing words.

The development of human language itself, 100,000 years ago or more, was (to flatter our own species) an information-processing revolution of the first rank. The fossil record suggests that the development of language was accompanied and furthered by the relatively rapid evolution of parts of the brain that specialize in language processing. We can think of this new neural circuitry, together with the accompanying development of the vocal cords, as a naturally evolved "technology," or mechanism, that makes language possible. This additional neural technology is apparently what gave rise to the remarkable universality of human speech—the ability to express in one language more or less what has been said in another. At the very least, language allowed the uniquely human forms of social organization that have made our species so successful thus far.

The farther back we go, the more momentous the information-processing revolutions we uncover. The development of the brain and the central nervous system was a triumph of naturally evolved technology, well suited for the transformation of information from the outside in, and for communication between parts of an organism. The development of multicellular organisms in the first place arose from numerous break-throughs in intra- and intercellular communication. Every successful mutation, every instance of speciation, constitutes an advance in information processing. But for an even greater revolution, dwarfing all that followed, we turn the clock back a billion years, to the invention of sex.

The original sexual revolution was a tour de force, a huge success that came from what at first glance looks like a bad idea. Why bad? Because it risked losing valuable information. A successful bacterium, reproducing *asexually*, passes on its exact genetic makeup (absent the occasional mutation) to its offspring. But if an organism reproduces *sexually*, its genes are scrambled with those of its mate in order to produce the off-spring's genes, a process called recombination. Because each half of this offspring's genes came from a different parent, and because of the scrambling process, no matter how successful either parent's unique combination of genes, the offspring's genome will not be the same as that parent's. Sexual reproduction has never passed on a full winning combination intact. Sex messes with success.

So why is sex good? From the point of view of natural selection, it allows for greater genetic variation at the same time as it faithfully repro-duces individual genes. Suppose the world were to get hotter. A hereto-fore successful, but asexual, bacterium would suddenly find itself in a hostile environment. Its previously well-adapted, almost perfectly iden-tical offspring would now be poorly adapted.

Without sex, the only way for bacteria to adapt is through mutation, which is caused by reproductive error or environmental damage. Most mutations are hurtful; they make for even less successful bacteria—though eventually, with luck, a mutation would arise that made for a more heat-resistant bacterium. Asexual adaptation is problematic because the dictate of the world, "Change or die," runs directly counter to one of the primary dictates of life: "Maintain the integrity of the genome." In engineering, this type of clash is called a coupled design. Two functions of a system clash so that it is not possible to adjust one without negatively affecting the other. In sexual reproduction, by contrast, the inherent scrambling, or recombination, affords a vast scope for change, yet still maintains genetic integrity.

Consider a small town with 1,000 inhabitants. Count up all the possible mating combinations (judging from daytime TV, there are quite a few), and then the number of possible ways their genes can be scrambled up and recombined in their kids. The result: the town is a genetic powerhouse, capable of generating as much diversity as billions of bacteria. This diversity is good: if an epidemic hits the town, there are likely to be survivors, who will then pass on their resistant genes to their children. Moreover, the capacity for diversity that sex conveys now comes without damage to the genome. By separating the function of adaptation from the function of maintaining the integrity of individual genes, sex allows much greater diversity while still keeping genes whole. Sex is not only fun, it is good engineering practice.

Moving even farther back in time, we come to the grandmother of all information-processing revolutions, life itself. About one-third of the way back to the beginning of the universe, life began on Earth. (When or whether it happened elsewhere is not known.) Living organisms possess genes, sequences of atoms in molecules such as DNA that encode information. The amount of information in a gene can be measured: the human genome possesses some 6 billion bits of information. Organisms pass their genetic information on to their offspring, sometimes in a mutated form. The organisms that are good at passing on genetic information are by definition successful; the organisms that fail to pass on their genes die out. Genetic information that conveys a reproductive advantage to its host tends to persist over generations, even as the organisms that carry it are born, reproduce, and die.

As it is passed down, genetic information is transmitted through natural selection. Genes and the mechanisms for copying and reproducing them are the key information-processing technology of life. Not surprisingly, the sum total of all genetic information processing performed by living organisms dwarfs the information processing performed by man-made computers, and should continue to do so for quite some time.

Surely, life is the big one. What revolution could top the origins of life in sheer power and beauty? But there was indeed an earlier information-processing revolution, one whose consequences encompass everything. The original information processor is the universe itself. Every atom, every elementary particle registers information. Every collision between atoms, every dynamic change in the universe, no matter how small, processes that information in a systematic fashion.

This computational capacity of the universe underlies all subsequent information-processing revolutions. Once a physical system possesses the ability to process information at a rudimentary level—performing simple operations on a few bits at a time—arbitrarily complicated forms of information processing can be built up from these basic operations. The laws of physics allow simple information processing at the quantum-mechanical level: one particle, one bit; one bump, one op. The complex forms of information processing we see around us—life, reproduction, language, society, video games—are all built up from simple operations governed by the laws of physics and performed on a few quantum bits at a time. Every information-processing revolution is associated with a new technology—the computer, the book, the brain, DNA. These technologies allow information to be registered and processed according to a set of rules. What is the technology associated with the Big Bang's information processing? What machine processes information in the computational universe? To see this universal information-processing technology in action, one need only open one's eyes and look around. The machine performing the "universal" computation is the universe itself.

CHAPTER 2

Computation

Information

I began the initial meeting of my MIT graduate course on information in the manner I begin all of my courses: "First," I said to the twenty-odd students, "you ask questions and I'll try to answer them. Second, if you don't ask questions, I'll ask you questions. Third, if you don't answer my questions, I'll tell you something I think you ought to know. Any questions?"

I waited. No response.

Something was wrong. Normally, MIT students are more than happy to try to stump the professor, particularly if the alternative is that the professor will try to stump them.

I moved on to step two: "No questions? Then here's one for you: What is information?"

Nothing. This was even worse. After all, these students had been stuffing themselves full of information since freshman year. If they didn't regurgitate some of it, I was going to have to resort to step three.

"OK. How about this one: What is the *unit* of information?"

At once, the class responded, "The bit!"

What do my students' answers, or lack thereof, reveal? That it is far easier to measure a quantity of information than to say what information is. And more broadly, "How much?" is frequently an easier question to answer than "What is . . . ?" What is energy? What is money? These are hard questions. How much energy does it take to . . . ? How much money does it take to . . . ? These questions have precise and available answers.

"What is a bit?" I asked. Now replies came thick and overlapping: "0 or 1!" "Heads or tails!" "Yes or no!" "True or false!" "The choice between two alternatives!"

All of these answers were correct. The word "bit" stands for "binary digit." "Binary" means consisting of two parts, and a bit represents one of these two alternatives. Traditionally, these alternatives are referred to as 0 and 1, but any two distinct alternatives (hot/cold, black/white, in/out) register a bit.

A bit is the smallest unit of information. A coin toss yields one bit: heads or tails. Two bits represent a slightly larger chunk of information. Two coin tosses yield one of four (or two times two) alternative outcomes: heads-heads, heads-tails, tails-heads, tails-tails. Similarly, three coin tosses yield one of eight (or two times two times two) alternatives.

As you can see from even these few examples, as you keep on tossing coins, the number of total alternatives—total possible outcomes of the series of tosses—grows rapidly. In fact, with each subsequent toss (remember: each toss yields one bit), the number of total alternatives doubles. So, to calculate the number of alternative outcomes in a given scenario, you simply multiply two by itself a number of times equal to the number of bits. For example, ten bits gives two multiplied by itself ten times, or 1,024 alternatives ($2 \times 2 \times 2 \times 2 \times 2 \times 2 \times 2 \times 2 \times 2 \times 2 = 2^{10} = 1,024 \approx 10^3$).

To put it in another way, ten bits correspond to about three digits, meaning a place in the "ones," the "tens," and the "hundreds" column as we traditionally count. Measuring amounts of information is simply a matter of counting. Counting in terms of bits is simpler, though less familiar to most, than counting in terms of digits. Counting digits from 0 to 9 is straightforward: 0, 1, 2, 3, 4, 5, 6, 7, 8, 9. At this point, though, you have run out of digits, so the next number is written 1 followed by 0, or 10. The number 10 has a 1 in the tens column and a 0 in the ones column. The next number, 11, has a 1 in the tens column and a 1 in the ones column. You can keep on counting in this vein up to 99. The next number is 100, with a 1 in the hundreds column, a 0 in the tens column, and a 0 in the ones column. (You can see why it wasn't so easy to grasp this way of counting the first time around, when you were five or so.)

Counting with bits works in a similar fashion. Start counting: $0 =$ zero, $1 =$ one. So far, so good, but now we have run out of bits. The next combination of bits is 10, which equals two: that is, a 1 in the "twos" column and a 0 in the "ones" column. (The representation of two as "10" is the feature of binary arithmetic that causes the first-time user the most trouble, as in "There are 10 kinds of people: those who know binary, and those who don't.") The next combination is 11, which equals three: a 1 in the twos column and a 1 in the ones column. Now we've run out of two-bit numbers.

The next combination is 100, which equals four: a 1 in the fours column, a 0 in the twos column, and a 0 in the ones column. Then comes 101, which equals five (1 in the fours column plus 1 in the ones column), $110 =$ six, $111 =$ seven. Eight is represented by four bits: 1000, with a 1 in the eights column and 0s in the fours, twos, and ones columns. Because they have only two bits instead of ten digits, the binary numbers get longer more quickly than ordinary, digital numbers.

Just as the powers of ten (tens, hundreds, thousands, millions) are important numbers in the ordinary, decimal way of counting, the powers of two are important numbers for counting with bits: $1 =$ one $= 2^0$, $10 =$ two $= 2^1$, $100 =$ four $= 2^2$, $1000 =$ eight $= 2^3$, $10000 =$ sixteen $= 2^4$, $100000 =$ thirty-two $= 2^5$, $1000000 =$ sixty-four $= 2^6$, $10000000 =$ one hundred and twenty-eight $= 2^7$. These numbers should look familiar to cooks. The English system of weights and measures is a binary system: eight ounces in a cup, sixteen in a pint (the American pint, that is—as in "a pint's a pound the world around"; the imperial pint is twenty ounces, and the troy pint is twelve ounces), thirty-two in a quart, sixty-four in a half gallon, and one hundred and twenty-eight in a gallon. Expressing numbers in binary notation is no more difficult than expressing measures in quarts, pints, and cups. One hundred and forty-six ounces, for example, is one gallon plus one pint plus one quarter cup: 128 + 16 + 2 = 146. Written in binary, 146 is equal to 10010010 with a one in the "gallons" column, a one in the "pints" column, a one in the "quarter cups" column, with zeros everywhere else. To translate a number into binary, just measure it out in teaspoons.

Figure 2. Counting in Binary

The American system of measuring volume is a binary system.

One hundred and forty-six ounces is one gallon plus one pint plus one quarter cup. In binary notation, 146 = 10010010.

Just as counting in binary is simple (though perhaps not easy for those of us who are new to it), so is binary arithmetic. The entire binary addition table consists of $0 + 0 = 0$, $0 + 1 = 1$, $1 + 1 = 10$. The binary multiplication table is even simpler: $0 \times 0 = 0$, $0 \times 1 = 0$, $1 \times 1 = 1$. Binary is beautiful.

Binary is also practical. The compact nature of binary notation makes it easy to construct simple electronic circuits to do binary arithmetic. These circuits, in turn, are the basis for digital computers. We may not be able to define information, but we can certainly use it.

Precision

"What if there are an infinite number of alternatives?" a student asked. "For example, there are an infinite number of real numbers between 0 and 1."

"If you have an infinite number of alternatives, then you have an infinite amount of information," I replied. Pick a binary number: 1001001 0110110 0100000 1110100 1101000 1100101 0100000 1100010 1100101 1100111 1101001 1101110 1101110 1101001 1101110 1101111, for example. Under the common coding scheme known as ASCII (American Standard Code for Information Interchange), each letter or typewriter character is given a seven-bit codeword.

Interpreted in ASCII, this number corresponds to the characters I = 1001001, n = 1101110, (SPACE) = 0100000, t = 1110100, h = 1101000, e = 1100101, (SPACE) = 0100000, b = 1100010, e = 1100101, g = 1100111, i = 1101001, n = 1101110, n = 1101110, i = 1101001, n = 1101110, g = 1100111—that is, the beginning of the passage "In the beginning was the word . . ." By adding more bits, you can produce a number corresponding to the entire text of the Gospel according to John. By adding yet more, you can get the rest of the Bible, followed by the Koran, followed by the Lotus Sutra, followed by all the books in the Library of Congress, and so on. An infinite number of alternatives corresponds to an infinite number of digits or bits—in other words, to an infinite amount of information.

In practice, however, the number of alternatives in any finite system is finite, so the amount of information is finite, too. Normally, we think of

5/2069138

quantities such as length, height, and weight as varying continuously: just as there are an infinite number of real numbers between 0 and 1, there are apparently an infinite number of possible lengths between zero meters and one meter. The reason that apparently continuous quantities such as the length of a metal rod can register only a finite amount of information is that these quantities are typically defined only to a finite level of precision. To see the trade-off between precision and information, think of measuring the length of that rod using a meterstick. The meterstick is made of wood. One hundred centimeters are marked and numbered on the stick. One thousand millimeters are marked, ten for each centimeter, but there is not enough room on the meterstick to number them legibly. You can use the meterstick to measure the length of the rod to the accuracy of about a millimeter. Below a millimeter, a meterstick does not measure distances well, simply because its physical characteristics give it a finite resolution. The total number of alternatives is 1,000, corresponding to three digits of accuracy, or about ten bits of information.

A particularly famous metal rod is the one made of an alloy of platinum and iridium that sits in the International Bureau of Weights and Measures in Paris and that for almost a century defined the length of a meter. A meter was originally conceived as being one ten-millionth of the distance from the North Pole to the equator as measured along the Paris meridian. Our meterstick would measure this rod to be one meter long, plus or minus half a millimeter.

With a more accurate measuring device than a meterstick, more bits of information about the rod's length become available. Look at the rod under an optical microscope. Features can now be discerned down to a size on the order of the wavelength of visible light—slightly less than a micron, which is a millionth of a meter. A microscope applied to a meterstick could be used to measure the length of the rod to an accuracy of a micron. The microscope measures the rod to six digits of accuracy, corresponding to about twenty bits of information. A similar degree of accuracy can be obtained using an interferometer, a device that measures length in terms of the number of wavelengths of light. An interferometer that used light of wavelength one micron would measure the rod as one million wavelengths long.

More extreme means allow even greater degrees of accuracy. In principle one could take a device called an atomic force microscope, which images individual atoms on surfaces, and drag it along the rod, measuring the rod by the number of atoms along its length. The distance between atoms is on the order of one ten-billionth of a meter (10^{-10} meters), a distance known as an angstrom. Now we have ten digits of accuracy, or about thirty-three bits of information about the length of the rod.

Greater precision in measuring the length of a macroscopic object such as a rod is hard to get. In certain cases, it is possible to measure distances to a much higher degree of precision—for example, in physicist Norman Ramsey's experiments to measure charge separations as small as one billion billion billionth of a meter within a neutron. The total number of values a measurement device can distinguish is given by the range of values (e.g., a meter) the device can register divided by the highest precision to which the device can measure (e.g., a millimeter). Range divided by precision tells you how many distinguishable values the device can register. The amount of available information is given by the number of bits required to count the available values. A device that gets thirty-three bits of information (ten digits) about a continuous variable is doing very well indeed.

To get thirty-three bits of information about the length of our rod, we have to count that length in atoms: heroic amounts of effort are typically required to wring more than a few tens of bits of information out of a single continuous quantity such as the length of a rod. By contrast, if we use many individual quantities to register information, we can rapidly accumulate many bits. In a quantum computer, each atom registers a bit; to get thirty-three bits requires thirty-three atoms. Our rod contains something like a billion billion billion atoms. If each one registers a bit, then the atoms in the rod can register a billion billion billion bits, far more than the length of the rod on its own can register. In general, the best way to get more information is not to increase the precision of measurements on a continuous quantity, but rather to put together measurements on more and more quantities, each one of which may register only a few bits. This compiling of bits—or digital representation—is effective because the number of total alternatives described grows much faster than does the number of bits.

Recall the king in the folktale who foolishly agreed to reward the hero in grains of wheat: one grain for the first square of the chessboard, two grains for the second, four for the third, and so on up to two raised to the sixty-fourth power (2^{64}) grains for the last square. The total number of grains is thus 10 billion billion. If each was only a millimeter long, they would fill almost forty cubic kilometers. As this example illustrates, only a few bits are required to label one of a very large number of alternatives. To give each of the grains on the chessboard a unique barcode, for instance, would require only sixty-five bits, or sixty-five pieces of information. With only 300 bits, you could assign a unique barcode to each of the 10^{90} elementary particles in the universe. The astronomically huge number of possible genetic codes is the source of the incredible diversity of living things, but the information that produces those codes can be stored on a tiny chromosome.

Meaning

"But doesn't information have to mean something?" asked a student, troubled.

"You're right that when we think of information we normally associate it with meaning," I answered. "But the meaning of 'meaning' is not clear."

For thousands of years, philosophers have tried to determine what "meaning" means, with mixed success. The reason it is so hard to pin down is that the meaning of a piece of information depends very much on how the information is to be interpreted. If you don't know how a message is to be interpreted, then you don't know its meaning. For example, if I say to you, "Yes," when you haven't asked a question, then you don't know what I mean by "Yes." If you ask, "Can I have another piece of cake?" and I say, "Yes," then you know what I mean. If you ask, "What is two plus two?" and I say, "Yes," then you don't know what I mean (though you might start to suspect that I have only one answer to any question). If you ask, "What is two plus two?" and I say "Four," then you know what I mean. Meaning is a bit like pornography: you know it when you see it.

Consider the string of bits given earlier: 1001001 1101110 0100000 1110100 1101000 1100101 0100000 1100010 1100101 1100111 1101001 1101110 1101110 1101001 1101110 1100111. Interpreted as a message encoded in ASCII, this string means "In the beginning." But taken on its own, with no specification of how it is to be interpreted, it means nothing other than itself. Meaning is defined only relative to a scheme of interpretation, as the following conversation between Alice and Humpty Dumpty reveals:

"I don't know what you mean by 'glory,' " Alice said.

Humpty Dumpty smiled contemptuously. "Of course you don't—till I tell you. I meant 'there's a nice knock-down argument for you!' "

"But 'glory' doesn't mean 'a nice knock-down argument,' " Alice objected.

"When *I* use a word," Humpty Dumpty said, in a rather scornful tone, "it means just what I choose it to mean—neither more nor less."

"The question is," said Alice, "whether you *can* make words mean so many different things."

"The question is," said Humpty Dumpty, "which is to be master—that's all."

Lewis Carroll, the author of *Alice's Adventures in Wonderland* and *Through the Looking-Glass,* was in real life Charles Dodgson, the nominalist philosopher. Dodgson was quite happy with the notion of words meaning what he chose them to mean.

A common way to express the dependence of meaning on interpretation is to use Ludwig Wittgenstein's notion of a language game. This is a game in which words are assigned meanings in terms of the actions they elicit from the players. Wittgenstein's *Philosophical Investigations,* for example, begins with a simple language game: A builder can ask her assistant for either a block, a pillar, a slab, or a beam. If she says "Block," her assistant hands her a block. If she says "Slab," her assistant hands her a slab. In this simplest of language games, we infer that the assistant knows what the builder means when she says "Block": she means "Hand me the block."

As language games become more complicated, the meaning of meaning becomes less easy to tease out of the dynamics of the game. Part of the problem is that natural human language is frequently ambiguous: a given statement can often have many possible meanings. Part of the problem is that we do not fully understand how the brain responds to language, so that even if we know that "Block" means "Hand me the block," we don't know the physical mechanism by which the brain of the listener arrives at this meaning. It would be useful to have an example of a situation in which a piece of information can be interpreted in only one way and in which the mechanism eliciting a response is completely known.

Computers supply one such mechanism. Computers respond to languages called computer languages (Java, C, Fortran, BASIC). Such languages consist of simple commands—such as PRINT or ADD—that can be strung together to instruct the computer to perform complicated tasks. If you adopt Wittgenstein's perspective that the meaning of a piece of information is to be found in the action this information provokes, the meaning of a computer program written in a particular computer language is to be found in the actions the computer performs as it interprets that program. All the computer is doing is performing sequences of elementary logic operations, such as AND, NOT, and COPY (to be discussed further later). The computer program unambiguously instructs the computer to perform a particular sequence of those operations. The "meaning" of a computer program is thus universal, in the sense that two computers following the same instructions will perform the same set of information-processing operations and obtain the same result.

The unambiguous nature of a computer program means that one and only one meaning is assigned to each statement. If a statement in a computer language has more than one possible interpretation, an error message is the result: for computers, ambiguity is a bug. By comparison, human languages are rich in ambiguity: except in special circumstances, most statements in, for example, English, have a variety of potential meanings, and this is a key aspect of poetry, fiction, flirting, and plain everyday conversation. The ambiguity of human language is not a bug, it's a bonus!

Although meaning is hard to define, it is one of the most powerful features of information. The basic idea of information is that one physical system—a digit, a letter, a word, a sentence—can be put into correspondence with another physical system. The information stands for the thing. Two fingers can be used to represent two cows, two people, two mountains, two ideas. A word can stand for anything (anything for which we have a word): orange, cow, money, freedom. By putting words in sentences, one can express—well, anything that can be expressed in words. The words in sequence can stand for a complicated thought.

In the same way that words can represent ideas and things, so can bits. The word and the bit are means by which information is conveyed, though the interpreter must supply the meaning.

The Computer

"What is a computer?" I asked my class. No answer. This was strange seeing as I was pretty sure my students had been using them since their first birthdays.

I waited. Finally someone volunteered an answer: "A machine that manipulates data stored as 0's and 1's." Another student disagreed, "You're talking about a *digital* computer. What about an analog computer? They store information on continuous voltage signals."

Eventually, everyone agreed on a significantly broader definition: a computer is a machine that processes information.

OK, I said, then what was the first computer? Now the class had warmed up: "The Mark 1." "Babbage's mechanical computer." "The slide rule." "The abacus." "The brain." "DNA."

A raised hand wiggled its fingers: "Digits!"

Clearly, if you define a computer as a machine that processes information, then pretty much anything can compute.

"For the moment," I said, "let's just look at machines that people fabricate to process information and save the question of people as information-processing machines for later."

Computers date back to the early days of *Homo sapiens*. Like the first tools, the first computers were rocks. "Calculus" is the Latin word for pebble, and the first calculations were performed by arranging and rearranging just that. And rock computers didn't have to be small. Stonehenge may well have been a big rock computer for calculating the relations between the calendar and the arrangement of the planets.

The technology used in computing puts intrinsic limits on the computations that can be performed (think rocks versus an Intel Pentium IV). Rock computers are good for counting, adding, and subtracting, but they aren't so good at multiplying and dividing. And to deal with large numbers, you need a lot of rocks. Several thousand years ago, someone had the bright idea of combining rocks with wood: If you kept the rocks in grooves on a wooden table, they were easier to move back and forth. Then it was discovered that if you used beads instead of rocks and mounted them on wooden rods, the beads were not only easy to move back and forth but also hard to lose.

The wooden computer, or abacus, is a powerful calculating tool. Before the invention of electronic computers, a trained abacus operator could out-calculate a trained adding-machine operator. But the abacus is not merely a convenient machine for manipulating pebbles. It embodies a powerful mathematical abstraction: zero. The concept of zero is the crucial piece of the Arabic number system—a system allowing arbitrarily large numbers to be represented and manipulated with ease—and the abacus is its mechanical incorporation. But which came first? Given the origin of the word for zero and the antiquity of the first abacus, it seems likely that the machine did.* Sometimes, machines make ideas.

Ideas also make machines. First rock, then wood: what material would supply the next advance in information processing? Bone. In the early seventeenth century, the Scottish mathematician John Napier discovered a way of changing the process of multiplication into addition. He carved ivory into bars, ruled marks corresponding to numbers on the bars, and then performed multiplication by sliding the bars alongside each other until the marks corresponding to the two numbers lined up. The total length of the two bars together then gave the product of the two numbers. The slide rule was born.

In the beginning of the nineteenth century, an eccentric Englishman named Charles Babbage proposed making computers out of metal. Babbage's "Difference Engine," which was meant to calculate various trigonometric and logarithmic tables, was to be built out of gears and shafts, like a steam engine. Each gear would register information by its position, and then process that information by meshing with other gears and turning. Though entirely mechanical in its construction, the way Babbage's machine organized information foreshadowed the way contemporary electronic computers organize information. It was designed with a central processing unit and a memory bank that could hold both program and data.

Despite generous financial support from the British crown, Babbage's

---

*By 1700 B.C. the Babylonians had a well-established "Arabic" number system, but zero was inferred from context, not written (i.e., 210 and 21 were written the same way). The oldest known "proto-abacus," the Salamis counting tablet, dates to 300 B.C. The use of 0 for zero was introduced by Ptolemy in A.D. 130 and was well established in India by A.D. 650.

plan to build a Difference Engine failed. Early-nineteenth-century technology possessed neither sufficiently precise machining techniques nor sufficiently hard alloys to construct the engine's gears and shafts. (The effort wasn't wasted, however: the development by Babbage's technicians of more precise machining techniques and harder alloys significantly accelerated the ongoing Industrial Revolution.) Although effective mechanical calculators were available by the end of the nineteenth century, large-scale working computers had to await the development of electronic circuit technology in the beginning of the twentieth.

By 1940, an international competition had arisen between various groups to construct computers using electronic switches such as vacuum tubes or electromechanical relays. The first simple electronic computer was built by Konrad Zuse in Germany in 1941, followed by large-scale computers in the United States and Great Britain later in the 1940s. These consisted of several rooms' worth of vacuum tubes, switching circuits, and power supplies, but they were puny in terms of their computational power—about a million times less powerful than the computer on which this book is being written.

Though costly, the initial electronic computers were useful enough that efforts were made to streamline them. In the 1960s, vacuum tubes and electromechanical relays were replaced by transistors, semiconductor switches that were smaller and more reliable and required less energy. A semiconductor is a material such as silicon that conducts electricity better than insulators such as glass or rubber, but less well than conductors such as copper. Starting in the late 1960s, the transistors were made still smaller by etching them on silicon-based integrated circuits, which collected all the components needed to process information on one semiconductor chip.

Since the 1960s, advances in photolithography—the science of engineering ever more detailed circuits—have halved the size of the components of integrated circuits every eighteen months or so. As a result, computers have doubled in power at the same rate, the phenomenon known as Moore's law. Nowadays, the wires in the integrated circuits in a run-of-the-mill computer are only 1,000 atoms wide.

For future reference, let me define some of the types of computers to which I will refer. A digital computer is a computer that operates by applying logic gates to bits; a digital computer can be electronic or mechanical. A classical computer is a computer that computes using the laws of classical mechanics. A classical digital computer is one that computes by performing classical logical operations on classical bits. An electronic computer is one that computes using electronic devices such as vacuum tubes or transistors. A digital electronic computer is a digital computer that operates electronically. An analog computer is one that operates on continuous signals as opposed to bits; it gets its name because such a computer is typically used to construct a computational "analog" of a physical system. Analog computers can be electronic or mechanical. A quantum computer is one that operates using the laws of quantum mechanics. Quantum computers have both digital and analog aspects.

Logic Circuits

What are our ever more powerful computers doing? They are processing information by breaking it up into its component bits and operating on those bits a few at a time. As noted earlier, the information to be processed is presented to the computer in the form of a program, a set of instructions in a computer language. The program is encoded in the computer's memory as a sequence of bits. For example, the command PRINT is written in ASCII code as P = 1010000 R = 1010010 I = 1001001 N = 1001110 T = 1010100. The computer looks at the program a few bits at a time, interprets the bits as an instruction, and executes the instruction. Then it looks at the next few bits and executes that instruction. And so on. Complicated procedures can be built up from sets of simple instructions—but there's more on that to come.

Conventional computers consist primarily of electronic circuits that physically implement "logic circuits." Logic circuits allow a complicated logical expression to be built up out of simple operations that act on just a few bits at a time. Physically, logic circuits consist of bits, wires, and gates. Bits, as we have seen, can register either 0 or 1; wires move the bits from one place to another; gates transform the bits one or two at a time.

For example, a NOT gate takes its input bit and flips it: that is, NOT transforms 0 into 1 and 1 into 0. A COPY gate makes a copy of a bit: it transforms an input bit 0 into two output bits 00 and an input bit 1 into two output bits 11. An AND gate takes two input bits and produces a single output bit equal to 1 if, and only if, both input bits are equal to 1; otherwise it produces the output 0. An OR gate takes two input bits and produces an output bit equal to 1 if one or both of the input bits is equal to 1; if both input bits are equal to 0, then it produces the output 0. Since the 1854 publication of *An Investigation of the Laws of Thought* by the logician George Boole of Queen's College, Cork, it has been known that

Figure 3. Logic Gates

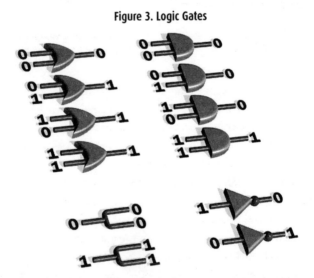

Logic gates are devices that take one or more input bits and transform them into one or more output bits. Clockwise from upper left: OR gate, AND gate, NOT gate, and COPY gate.

any desired logical expression, including complex mathematical calculations, can be built up out of NOT, COPY, AND, and OR. They make up a universal set of logic gates.

Boole's Laws of Thought imply that any logical expression or computation can be encoded as a logic circuit. A digital computer is a computer that operates by implementing a large logic circuit consisting of millions of logic gates. Familiar computers such as Macintoshes and PCs are electronic realizations of digital computers.

Figure 4. Logic Circuits

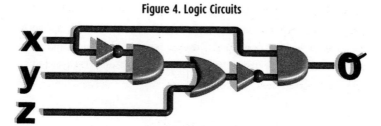

AND, OR, NOT, and COPY gates can be wired together to make logic circuits. A logic circuit can perform more complicated transformations of its input bits.

In an electronic computer, bits are registered by electronic devices such as capacitors. A capacitor is like a bucket that holds electrons. To fill the bucket, a voltage is applied to the capacitor. A capacitor at zero voltage has no excess electrons and is said to be uncharged. An uncharged capacitor in a computer registers a 0. A capacitor at non-zero voltage holds lots of excess electrons and registers a 1.

Capacitors are not the only electronic devices that computers use to store information. In your computer's hard drive, bits are registered by tiny magnets: a magnet whose north pole points up registers a 0 and a magnet whose north pole points down registers a 1. As always, any device that has two reliably distinguishable states can register a bit.

In a conventional digital electronic computer, logic gates are implemented using transistors. A transistor can be thought of as a switch. When the switch is open, current can't flow through it. When the switch is closed, current flows through. A transistor has two inputs and one output. In an *n-type* transistor, when the first input is kept at a low voltage the switch is open and current can't flow from the second input to the output; raising the voltage on the first input allows current to flow. In a *p-type* transistor, when the first input is kept at low voltage the switch is closed, so current can flow from the second input to the output; *n-* and *p-type* transistors can be wired together to create AND, OR, NOT, and COPY gates.

When a computer computes, all it is doing is applying logic gates to bits. Computer games, word processing, number crunching, and spam all arise out of the electronic transformation of bits, one or two at a time.

Uncomputability

Up to this point, we have emphasized the underlying simplicity of information and computation. A bit is a simple thing; a computer is a simple machine. But this doesn't mean that computers are incapable of complex and sophisticated behavior. One counterintuitive result of a computer's fundamentally logical operation is that its future behavior is intrinsically *un*predictable. The only way to find out what a computer will do once it has embarked upon a computation is to wait and see what happens.

In the 1930s, the Austrian logician Kurt Gödel showed that *any* sufficiently powerful mathematical theory has statements that, if false, would render the theory inconsistent but that cannot be proved to be true. All sufficiently powerful systems of logic contain unprovable statements. The computational analog of an unprovable statement is an uncomputable quantity.

A well-known problem whose answer is uncomputable is the so-called halting problem: Program a computer. Set it running. Does the computer ever halt and give an output? Or does it run forever? There is no general procedure to compute the answer to this question. That is, no computer program can take as input another computer program and determine with 100 percent probability whether the first computer program halts or not.

Of course, for many programs, you *can* tell whether or not the computer will halt. For example, the program PRINT 1,000,000,000 clearly halts: A computer given this program as input prints 1,000,000,000, then halts. But as a rule, no matter how long a computer has gone on computing without halting, you cannot conclude that it will never halt.

Although it may sound abstract, the halting problem has many practical consequences. Take, for example, the debugging of computer programs. Most computer programs contain "bugs" or errors that make the computer behave in unexpected ways, e.g., crash. It would be useful to have a "universal debugger" for computer programs. Such a debugger would take as input the computer program, together with a description of what the program is supposed to accomplish, and then check to see that the program does what it is supposed to do. Such a debugger cannot exist.

The universal debugger is supposed to verify that its input program gives the right output. So the first thing a universal debugger should check is whether the input program has any output at all. But to verify that the program gives an output, the universal debugger needs to solve the halting problem. That it cannot do. The only way to determine if the program will halt is to run it and see, and at that point, we no longer need the universal debugger. So the next time a bug freezes your computer, you can take solace in the deep mathematical truth that there is no systematic way to eliminate all bugs. Or you can just curse and reboot.

Gödel showed that the capacity for self-reference leads automatically to paradoxes in logic; the British mathematician Alan Turing showed that self-reference leads to uncomputability in computers. It is tempting to identify similar paradoxes in how human beings function. After all, human beings are masters of self-reference (some humans seem capable of no other form of reference) and are certainly subject to paradox.

Humans are notoriously unable to predict their own future actions. This is an important feature of what we call free will. "Free will" refers to our apparent freedom to make decisions. For example, when I sit down in a restaurant and look at the menu, I and only I decide what I will order, and before I decide, even I don't know what I will choose. That is, our own future choices are inscrutable to ourselves. (They may not, of course, be inscrutable to others. For years my wife and I would go for lunch to Josie's in Santa Fe. I, after spending a long time scrutinizing the menu, would always order the half plate of chiles rellenos, with red and green chile, and posole instead of rice. I felt strongly that I was exercising free will: until I chose the rellenos half plate, I felt anything was possible. My wife, however, knew exactly what I was going to order all the time.)

The inscrutable nature of our choices when we exercise free will is a close analog of the halting problem: once we set a train of thought in motion, we do not know whether it will lead anywhere at all. Even if it does lead somewhere, we don't know where that somewhere is until we get there.

Ironically, it is customary to assign our own unpredictable behavior and that of other humans to irrationality: were we to behave rationally, we reason, the world would be more predictable. In fact, it is just when we behave rationally, moving logically, like a computer, from step to step, that our behavior becomes provably *un*predictable. Rationality combines with self-reference to make our actions intrinsically paradoxical and uncertain.

This lovely inscrutability of pure reason harks back to an earlier account of the role of logic in the universe. From his home in Cordova, the twelfth-century Muslim philosopher Averroës (Ibn Rushd) in his studies of Aristotle concluded that what is immortal in human beings is not their soul but their capacity for reason. Reason is immortal exactly because it is not specific to any individual; instead, it is the common property of all reasoning beings.

Computers certainly possess the ability to reason and the capacity for self-reference. And just because they do, their actions are intrinsically inscrutable. Consequently, as they become more powerful and perform a more varied set of tasks, computers exhibit an unpredictability approaching that of human beings. Indeed, by Averroës's standards, they possess the same degree of immortality as humans.

Programming computers to perform simple human tasks is difficult: getting a computerized robot to vacuum a room or empty a dishwasher, even to minimal standards, is a problem that has outstripped the abilities of several generations of researchers in artificial intelligence. By contrast, no special effort is required to program a computer to behave in unpredictable and annoying ways. When it comes to their capacity to screw things up, computers are becoming more human every day.*

*As in the bumper sticker "To err is human. To really screw things up requires a computer."

CHAPTER 3

The Computational Universe

The Story of the Universe: Part One

The universe is made of atoms and elementary particles, such as electrons, photons, quarks, and neutrinos. Although we will soon delve into a vision of the universe based on a computational model, we would be foolish not to first explore the stunning discoveries of cosmology and elementary-particle physics. Science already affords us excellent ways of describing the universe in terms of physics, chemistry, and biology. The computational universe is not an alternative to the physical universe. The universe that evolves by processing information and the universe that evolves by the laws of physics are one and the same. The two descriptions, computational and physical, are complementary ways of capturing the same phenomena.

Of course, humans have been speculating about the origins of the universe far longer than they have been dabbling in modern science. Telling stories about the universe is as old as telling stories. In Norse mythology, the universe begins when a giant cow licks the gods out of the salty lip of a primordial pit. In Japanese mythology, Japan arises from the incestuous embrace of the brother and sister gods Izanagi and Izanami. In one Hindu creation myth, all creatures rise out of the clarified butter obtained from the sacrifice of the thousand-headed Purusha. More recently, though, over the last century or so, astrophysicists and cosmologists have constructed a detailed history of the universe supported by observational evidence.

The universe began a little less than 14 billion years ago, in a huge explosion called the Big Bang. As it expanded and cooled down, various familiar forms of matter condensed out of the cosmic soup. Three minutes after the Big Bang, the building blocks for simple atoms such as hydrogen and helium had formed. These building blocks clumped together under the influence of gravity to form the first stars and galaxies 200 million years after the Big Bang. Elements heavier than iron were produced when these early stars exploded in supernovae. Our own sun and solar system formed 5 billion years ago, and life on Earth was up and running a little over a billion years later.

This conventional history of the universe is not as sexy as some versions, and dairy products enter into only its later stages, but unlike older creation myths, the scientific one has the virtue of being consistent with known scientific laws and observations. And even though it is phrased in terms of physics, the conventional history of the universe still manages to make a pretty good story. It has drama and uncertainty, and many questions remain: How did life arise? Why is the universe so complex? What is the future of the universe and of life in particular? When we look into the Milky Way, our own galaxy, we see many stars much like our own. When we look beyond, we see many galaxies apparently much like the Milky Way. There is a scripted quality to what we see, in which the same astral dramas are played out again and again by different stellar actors in different places. If the universe is in fact infinite in extent, then somewhere, eventually, every possible scenario allowed by the laws of physics will be enacted. The story of the universe is a kind of cosmic soap opera whose actors play out all possible permutations of the drama.

Energy: The First Law of Thermodynamics

Let's get acquainted with the major players in the cosmic soap. In conventional cosmology, the primary actor is energy—the radiant energy in light and the mass energy in protons, neutrons, and electrons. What is energy? As you may have learned in middle school, energy is the ability to do work. Energy makes physical systems do things.

Famously, energy has the feature of being conserved: it can take different forms—heat, work, electrical energy, mechanical energy—but it is never lost. This is known as the first law of thermodynamics. But if energy is conserved, and if the universe started from nothing, then where did all of the energy come from? Physics provides an explanation.

Quantum mechanics describes energy in terms of quantum fields, a kind of underlying fabric of the universe, whose weave makes up the elementary particles—photons, electrons, quarks. The energy we see around us, then—in the form of Earth, stars, light, heat—was drawn out of the underlying quantum fields by the expansion of our universe. Gravity is an attractive force that pulls things together. (As high school students will tell you, "Gravity sucks.") As the universe expands (which it continues to do), gravity sucks energy out of the quantum fields. The energy in the quantum fields is almost always positive, and this positive energy is exactly balanced by the negative energy of gravitational attraction. As the expansion proceeds, more and more positive energy becomes available, in the form of matter and light—compensated for by the negative energy in the attractive force of the gravitational field.

The conventional history of the universe pays great attention to energy: How much is there? Where is it? What is it doing? By contrast, in the story of the universe told in this book, the primary actor in the physical history of the universe is *information*. Ultimately, information and energy play complementary roles in the universe: Energy makes physical systems do things. Information tells them what to do.

Entropy: The Second Law of Thermodynamics

If we could look at matter at the atomic scale, we would see atoms dancing and jiggling every which way at random. The energy that drives this random atomic dance is called heat, and the information that determines the steps of this dance is called *entropy*. More simply, entropy is the information required to specify the random motions of atoms and molecules—motions too small for us to see. *Entropy is the information contained in a physical system that is invisible to us.*

Entropy is a measure of the degree of molecular disorder existing in a system: it determines how much of the system's thermal energy is unavailable for conversion into mechanical work—how much of its energy is useful. The second law of thermodynamics states that the entropy of the universe as a whole does not decrease; in other words, the amount of unusable energy is increasing. Manifestations of the second law are all around us. Hot steam can run a turbine and do useful work. As steam cools, its jiggling molecules transfer some of their disorder into increased disorder in the surrounding air, heating it up. As the molecules of steam jiggle slower and slower, the air molecules jiggle faster and faster, until steam and air are at the same temperature. When the difference in temperatures is minimized, the entropy of the system is maximized. But room temperature steam will do no work.

Here is yet another way to conceive of entropy. Most information is invisible. The number of bits of information required to characterize the dance of atoms vastly outweighs the number of bits we can see or know. Consider a photograph: It has an intrinsic graininess determined by the size of the grains of silver halide that make up the photographic film—or, if it is a digital photograph, by the number of pixels that make up the digital image on a screen. A high-quality digital image can register almost a billion bits of visible information. How did I come up with that number? One thousand pixels per inch is a high resolution, comparable to the best resolution that can be distinguished with the naked eye. At this resolution, each square inch of the photograph contains a million pixels. An 8- by 6-inch color photograph with 1,000 pixels per inch has 48 million pixels. Each pixel has a color. Digital cameras typically use 24 bits to produce 16 million colors, comparable to the number that the human eye can distinguish. So an 8- by 6-inch color digital photograph with 1,000 pixels per inch and 24 bits of color resolution has 1,152,000,000 bits of information. (An easier way to see how many bits are required to register a photograph is to look at how rapidly the memory space in your digital camera disappears when you take a picture. A typical digital camera takes high-resolution pictures with 3 million bytes—3 megabytes—of information. A byte is 8 bits, so each picture on the digital camera registers approximately 24 million bits.)

1,152,000,000 bits is a lot of information, but the amount of information required to describe the invisible jiggling of the atoms in the grains of silver halide of a non-digital photograph is much greater. To describe them would require more than a million billion billion bits (10^{24}, or a 1 followed by 24 zeros). The invisible jiggling atoms register vastly more information than the visible photograph they make up. A photograph that registered the same amount of visible information as the invisible information in a gram of atoms would have to be as big as the state of Maine.

The number of bits registered by the jiggling atoms that make up the photographic image on film can be estimated as follows. A grain of silver halide is about a millionth of a meter across and contains about a trillion atoms. There are tens of billions of grains of silver halide in the photographic film. Describing where an individual atom (at room temperature) is in its infinitesimal dance requires 10 to 20 bits per atom. The total amount of information registered by the atoms in the photograph is thus 10^{23} bits. The billion (10^9) bits of information *visible* in the photograph, as represented by the digital image, represent only a tiny fraction of this total. The remainder of the information contained in the matter of the photograph is invisible. This invisible information is the entropy of the atoms.

Free Energy

The laws of thermodynamics guide the interplay between our two actors, energy and information. To experience another example of the first and second laws, take a bite of an apple. The sugars in the apple contain what is called free energy. Free energy is energy in a highly ordered form associated with a relatively low amount of entropy. In the case of the apple, the energy in sugar is stored not in the random jiggling of atoms but in the ordered chemical bonds that hold sugar together. It takes much less information to describe the form energy takes in a billion ordered chemical bonds than it does to describe that same energy spread among a billion jiggling atoms. The relatively small amount of information required to describe this energy makes it available for use: that's why it's called free.

Pick the apple, take a bite. You've ingested free energy. Your digestive system contains chemicals called enzymes that convert the apple's sugars into glucose, a form of sugar that can be used directly by your muscles. Every gram of glucose contains a few kilocalories of free energy. Once you've digested the sugar, you can run miles on a few hundred kilocalories. (A calorie is the amount of energy required to raise one gram of water one degree Celsius. A kilocalorie, 1,000 calories, is what someone on a diet would normally call a Calorie: a teaspoonful of sugar contains ten kilocalories of free energy. One hundred kilocalories is enough energy to lift a VW Bug one hundred feet in the air!) While you run, the free energy in the sugar is converted into motion by your muscles. By the time you're finished running, you're hot: the free energy in the sugar has been converted into heat and work. The number of calories of heat and work exactly matches the calories of free energy in the apple's sugar. In obedience to the first law of thermodynamics, the total amount of energy remains the same. (In obedience to the second law, the amount of information required to describe the extra jiggling of molecules in your hot muscles and sweaty skin is much greater than the amount of information required to describe the ordered chemical bonds in the apple's sugar.)

Unfortunately, to reverse this process is not so easy. If you wanted to convert the energy in heat, which has lots of invisible information (or entropy), back into energy in chemical bonds, which has much less entropy, you would have to do something with that extra information. As we will discuss, the problem of finding a place for the extra bits in heat puts fundamental limits on how well engines, humans, brains, DNA, and computers can function.

In either scenario, though, it's clear that energy and information (visible and invisible) are the two primary actors in the universal drama. The universe we see around us arises from the interplay between these two quantities, interplay governed by the first and second laws of thermodynamics. Energy is conserved. Information never decreases. It takes energy for a physical system to evolve from one state to another. That is, it takes energy to process information. The more energy that can be applied, the faster the physical transformation takes place and the faster the information is processed. The maximum rate at which a physical system can process information is proportional to its energy. The more energy, the faster the bits flip. Earth, air, fire, and water in the end are all made of energy, but the different forms they take are determined by information. To do anything requires energy. To specify what is done requires information. Energy and information are by nature (no pun intended) intertwined.

The Story of the Universe: Part Two

Now that our two protagonists have been introduced, let's tell the story of the universe in terms of their interplay. It is this interplay—this back-and-forth between information and energy—that makes the universe compute.

Over the last century, advances in the construction of telescopes have led to ever more precise observations of the universe beyond our solar system. The past decade has been a particularly remarkable one for observations of the heavens. Ground-based telescopes and satellite observatories have generated rich data describing what the universe looks like now, as well as what it looked like in the past. (Because the speed of light is finite, when you look at a galaxy that's a billion light-years away, you're looking at an image from a billion years ago.) The historical nature of cosmic observation proves useful as we attempt to untangle the early history of the universe.

The universe began just under 14 billion years ago in a massive explosion. What happened before the Big Bang? Nothing.* There was no time and no space. Not just empty space, but *the absence of space itself.* Time itself had a beginning. There is nothing wrong with beginning from nothing. For example, the positive numbers begin from zero (the "empty thing"). Before zero, there are no positive numbers. Before the Big Bang, there was nothing—no energy, no bits.

Then, all at once, the universe sprang into existence. Time began, and with it, space. The newborn universe was simple; the newly woven fabric of quantum fields contained only small amounts of information and energy. At most, it required a few bits of information to describe. In fact, if—as some physical theories speculate—there is only one possible initial state of the universe and only one self-consistent set of physical laws, then the initial state required *no* bits of information to describe. Recall that to generate information, there must be alternatives—e.g., 0 or 1, yes or no, this or that. If there were no alternatives to the initial state of the universe, then exactly zero bits of information were required to describe it; it registered zero bits. This initial paucity of information is consistent with the notion that the universe sprang from nothing.

As soon as it began, though, the universe began to expand. As it expanded, it pulled more and more energy out of the underlying quantum fabric of space and time. Current physical theories suggest that the amount of energy in the early universe grew very rapidly (a process called "inflation"), while the amount of information grew more slowly. The early universe remained simple and orderly: it could be described by

---

*In some cosmological theories the universe has been around forever; the Big Bang was preceded by a Big Crunch. In these models, our universe will expand, and then recontract in another Big Crunch, followed by another Big Bang, and so on. While allowed within the laws of physics, such oscillating universe models are not currently favored by observation.

just a few bits of information. The energy that was created was free energy.

This paucity of information did not last for long, however. As the expansion continued, the free energy in the quantum fields was converted into heat, increasing entropy, and all sorts of elementary particles were created. These particles were hot: they jiggled around with a vengeance. To describe this jiggling would take a lot of information. After a billionth of a second—the amount of time it takes light to travel about a foot—had passed, the amount of information contained within the universe was on the order of 100 million billion billion billion billion billion (10^{50}) bits. That's approximately one bit for every atom that makes up the Earth. To store that much information visually would require a photograph the size of the Milky Way. *The Big Bang was also a Bit Bang.*

As the energy in the universe changed its form, the universe also processed and transformed its bits, filling up its "memory register" with the results of this information processing. After that billionth of a second, the universe had performed about 10,000 billion billion billion billion billion billion billion billion (10^{67}) elementary operations, or "ops," on the bits it registered. A lot had happened. But what was the universe computing during this initial billionth of a second? Science fiction writers have speculated that entire civilizations could have arisen and declined during this time—a time very much shorter than the blink of an eye. We have no evidence of these fast-living folk. More likely, these early ops consisted of elementary particles bouncing off one another in random fashion.

After this first billionth of a second, the universe was very hot. Almost all of the energy that had been drawn into it was now in the form of heat. Lots of information would have been required to describe the infinitesimal jigglings of the elementary particles in this state. In fact, when all matter is at the same temperature, entropy is maximized. There was very little free energy—that is, *order*—at this stage, making the moments after the Big Bang a hostile time for processes like life. Life requires free energy. Even if there were some form of life that could have withstood the high temperatures of the Big Bang, that life-form would have had nothing to eat.

As the universe expanded, it cooled down. The elementary particles jiggled around more slowly. The amount of information required to describe their jiggles stayed almost the same, though, increasing gradually over time. It might seem that slower jiggles would require fewer bits to describe, and it's true that fewer bits were required to describe their *velocities*. But, at the same time, the amount of space in which they were jiggling was increasing, requiring more bits to describe their *positions*. Thus, the total amount of information remained constant or increased in accordance with the second law of thermodynamics.

As the jiggles got slower and slower, bits and pieces of the cosmic soup began to condense out. This condensation produced some of the familiar forms of matter we see today. When the amount of energy in a typical jiggle became less than the amount of energy required to hold together some form of composite particle—a proton, for example— those particles formed. When the jiggles of the constituent parts— quarks, in the case of a proton—were no longer sufficiently energetic to maintain them as distinct particles, they stuck together as a composite particle that condensed out of the cosmic soup. Every time a new ingredient of the soup condensed out, there was a burst of entropy—new information was written in the cosmic cookbook.

Particles condensed out of the jiggling soup in order of the energy required to bind them together. Protons and neutrons—the particles that make up the nuclei of atoms—condensed out a little more than a millionth of a second after the Big Bang, when the temperature was around 10 million million degrees Celsius. Atomic nuclei began to form at one second, and about a billion degrees. After three minutes, the nuclei of the lightweight atoms—hydrogen, helium, deuterium, lithium, beryllium, and boron—had condensed. Electrons were still whizzing around too fast, though, for these nuclei to capture them and form complete atoms. Three hundred eighty thousand years after the Big Bang, when the temperature of the universe had dropped to a little less than 10,000 degrees Celsius, electrons had finally cooled enough to be captured, and stable atoms formed.

Order from Chaos (the Butterfly Effect)

Until the formation of atoms, almost all the information in the universe lay at the level of the elementary particle. Nearly all bits were registered by the positions and velocities of protons, electrons, and so forth. On any larger scale, the universe still contained very little information: it was featureless and uniform. (How uniform was it? Imagine the surface of a lake on a windless morning so calm that the reflections of the trees are indistinguishable from the trees themselves. Imagine the earth with no mountain larger than a molehill. The early universe was more uniform still.)

Nowadays, by contrast, telescopes reveal huge variations and nonuniformities in the universe. Matter clusters together to form planets such as Earth and stars such as the sun. Planets and suns cluster together to form solar systems. Our solar system clusters together with billions of others to form a galaxy, the Milky Way. The Milky Way, in turn, is only one of tens of galaxies in a cluster of galaxies—and our cluster of galaxies is only one cluster in a supercluster. This hierarchy of clusters of matter, separated by cosmic voids, makes up the present, large-scale structure of the universe.

How did this large-scale structure come about? Where did the bits of information come from? These bits had their origins in the very early universe we've just explored. Their origins can be explained by the laws of quantum mechanics, coupled to the laws of gravity.

Quantum mechanics is the theory that describes how matter and energy behave at their most fundamental levels. At the small scale, quantum mechanics describes the behavior of molecules, atoms, and elementary particles. At larger scales, it describes the behavior of you and me. Larger still, it describes the behavior of the universe as a whole. *The laws*

of quantum mechanics are responsible for the emergence of detail and struc-
ture in the universe.

The theory of quantum mechanics gives rise to large-scale structure because of its intrinsically probabilistic nature. Counterintuitive as it may seem, quantum mechanics produces detail and structure because it is inherently uncertain.

The early universe was uniform: the density of energy was everywhere almost the same. But it was not *exactly* the same. In quantum mechanics, quantities such as position, velocity, and energy density do not have exact values. Instead, their values fluctuate. We can describe their probable values—the most likely location of a particle, for example—but we cannot claim perfect certainty. Because of these quantum fluctuations, some regions of the early universe were ever so slightly more dense than other regions.

As time passed, the attractive force of gravity caused more matter to move toward these denser regions, further increasing their energy density, and decreasing density in the surrounding volume. Gravity thus amplified the effect of an initially tiny discrepancy, causing it to increase. Just such a tiny quantum fluctuation near the beginning of time formed the seed for what would eventually become a cluster of galaxies. Slightly later on, further fluctuations formed the seeds for the positions of individual galaxies within the cluster, and still later fluctuations seeded the positions of planets and stars.

In the process of creating this large-scale structure, gravity also created the free energy that living things require to survive. As the matter clumped together, it moved faster and faster, gaining energy from the gravitational field; that is, the matter heated up. The larger the clump grew, the hotter the matter became. If enough matter clumped together, the temperature in the center of the clump rose to the point at which thermonuclear reactions are ignited: the sun began to shine! The light from the sun has lots of free energy—energy plants would use for photosynthesis, for example. As soon as plants came into existence, that is.

The ability of gravity to amplify small fluctuations in density is a reflection of a physical phenomenon known as "chaos." In a chaotic system, what begins as a tiny difference is amplified in time. Perhaps the most famous example of chaos is the so-called butterfly effect. The equations for motion within Earth's atmosphere are inherently chaotic; thus, a tiny perturbation, such as the flutter of a butterfly's wing, can be amplified over time and distance, becoming a hurricane months and miles down the line. The minuscule quantum fluctuations of energy density at the time of the Big Bang are the butterfly effects that would come to yield the large-scale structure of the universe.

Every galaxy, star, and planet owes its mass and position to quantum accidents of the early universe. But there's more: these accidents are also the source of the universe's minute details. Chance is a crucial element of the language of nature. Every roll of the quantum dice injects a few more bits of detail into the world. As these details accumulate, they form the seeds for all the variety of the universe. Every tree, branch, leaf, cell, and strand of DNA owes its particular form to some past toss of the quantum dice. Without the laws of quantum mechanics, the universe would still be featureless and bare. Gambling for money may be infernal, but betting on throws of the quantum dice is divine.

The Universal Computer

We have seen that the universe computes by registering and transforming information, so we might call what we see around us the "universal computer." But there is another, more technical meaning to that phrase. In computer science, a universal computer is a device that can be programmed to process bits of information in any desired way. Conventional digital computers of the sort on which this book is being written are universal computers, and their languages are universal languages. Human beings are capable of universal computation, and human languages are universal. Most systems that can be programmed to perform arbitrarily long sequences of simple transformations of information are universal.

Universal computers can do pretty much anything with information. Two of the inventors of universal computers and universal languages, Alonzo Church and Alan Turing, hypothesized that any possible mathematical manipulation can be performed on a universal computer; that is, universal computers can generate mathematical patterns of any level of complexity. A universal computer itself, though, need not be a complicated machine; all it must be able to do is take bits, one or two at a time, and perform simple operations upon them. Any desired transformation of however large a set of bits can be enacted by repeatedly performing operations on just one or two bits at a time. And any machine that can enact this sequence of simple logical operations is a universal computer.

Significantly, universal computers can be programmed to transform information in any way desired, and any universal computer can be programmed to transform information in the same way as any other universal computer. That is, any universal computer can simulate another, and vice versa. This intersimulatability means that all universal computers can perform the same set of tasks. This feature of computational universality is a familiar one: if a program will run on a PC, it can necessarily be translated to run on a Mac.

Of course, the program may take longer to run on the Mac than the PC, or vice versa. Programs written for a specific universal computer tend to run faster on that computer than the translated program runs on another. But the translated program will still run. In fact, every universal computer can be shown not only to simulate every other universal computer, but to do so *efficiently*. The slowdown due to translation is relatively insignificant.

Digital vs. Quantum

The universe computes. Its computer language consists of the laws of physics and their chemical and biological consequences. But is the universe nothing more than a universal digital computer, in the technical sense elucidated by Church and Turing? It is possible to give a precise scientific answer to this question. The answer is No.

The idea that the universe might be, at bottom, a digital computer is decades old. In the 1960s, Edward Fredkin, then a professor at MIT, and Konrad Zuse, who constructed the first electronic digital computers in Germany in the early 1940s, both proposed that the universe was fundamentally a universal digital computer. (More recently, this idea has found an advocate in the computer scientist Stephen Wolfram.) The idea is an appealing one: digital systems are simple, yet able to reproduce behavior of any degree of complexity. In particular, computers whose architecture mimics the structure of space and time (so-called cellular automata) can efficiently reproduce the motions of classical particles and the interactions between them.

In addition to the aesthetic appeal of a digital universe, there is powerful observational evidence for the computational ability of physical laws. The laws of physics clearly support universal computation. The problem with identifying the universe as a classical digital computer is that the universe appears to be significantly more computationally powerful.

Two computing machines have the same computational power if each can simulate the other efficiently. The key word here is "efficiently." The laws of physics can simulate digital computation efficiently; the universe effortlessly encompasses conventional digital computers. But now consider whether or not the universe can be simulated efficiently by a conventional computer. In fact, a conventional digital computer seems unable to simulate the universe efficiently.

At first, it might seem otherwise. After all, the laws of physics are apparently simple. Even if they turn out to be somewhat more complicated than we currently suspect, they are still mathematical laws that can be expressed in a conventional computer language; that is, a conventional computer can simulate the laws of physics and their consequences. If you had a large enough computer, then, you could program it (using, for example, a language such as Java) with descriptions of the initial state of the universe, and of the laws of physics, and set it running. Eventually, you would expect this computer to come up with accurate descriptions of the state of the universe at any later time.

The problem with such simulations is not that they are impossible, but that they are inefficient. The universe is fundamentally quantum-mechanical, and conventional digital computers have a hard time simulating quantum-mechanical systems. Why? Quantum mechanics is just as weird and counterintuitive for conventional computers as it is for human beings. In fact, in order to simulate even a tiny piece of the universe—consisting, say, of a few hundred atoms—for a tiny fraction of a second, a conventional computer would need more memory space than there are atoms in the universe as a whole, and would take more time to complete the task than the current age of the universe. Now *that's* inefficient.

This is not to say that classical computers are useless for capturing certain aspects of quantum behavior: they are quite good at calculating approximate energies and ground states of quantum systems. It's just that there is no known way for them to perform a full-blown dynamical simulation of a complex quantum system without using vast amounts of dynamical resources. Classical bits are very bad at storing the information required to characterize a quantum system: the number of bits grows *exponentially* with the number of pieces of the system. What does this mean? The failure of classical simulation of quantum systems suggests that the universe is intrinsically more computationally powerful than a classical digital computer.

But what about a *quantum* computer? A few years ago, acting on a suggestion from the physicist Richard Feynman, I showed that quantum computers can simulate any system that obeys the known laws of physics (and even those that obey as yet undiscovered laws!) in a straightforward and efficient way.

In brief, the simulation proceeds as follows: First, map the state of every piece of a quantum system—every atom, electron, or photon—onto the state of some small set of quantum bits, known as a quantum register. Because the register is itself quantum-mechanical, it has no problem storing the quantum information inherent in the original system on just a few quantum bits. Then enact the natural dynamics of the quantum system using simple quantum logic operations—interactions between quantum bits. Because the dynamics of physical systems consists of interactions between its constituent parts, these interactions can be simulated directly by quantum logic operations mapped onto the bits in the quantum register that correspond to those parts.

This method of quantum simulation is direct and efficient. The amount of time the quantum computer takes to perform the simulation is proportional to the time over which the simulated system evolves, and the amount of memory space required for the simulation is proportional to the number of subsystems or subvolumes of the simulated system. The simulation proceeds by a direct mapping of the dynamics of the system onto the dynamics of the quantum computer. Indeed, an observer that interacted with the quantum computer via a suitable interface would be unable to tell the difference between the quantum computer and the system itself. All measurements made on the computer would yield exactly the same results as the analogous measurements made on the system. Quantum computers, then, are universal quantum simulators.

The universe is a physical system. Thus, it could be simulated efficiently by a quantum computer—one exactly the same size as the universe itself. Because the universe supports quantum computation and can be efficiently simulated by a quantum computer, the universe is neither more nor less computationally powerful than a universal quantum computer.

In fact, the universe is *indistinguishable* from a quantum computer. Consider a quantum computer performing an efficient simulation of the universe. Now, compare the results of measurements taken in the universe with measurements taken in the quantum computer. Measurements in the universe are taken by one piece of the universe—in this case, us—on the remainder. The analogous processes occur in a quantum computer when one register of the computer gains information about another register. Because the quantum computer can perform an efficient and accurate simulation, the results of these two sets of measurements will be indistinguishable.

The universe possesses the same information-processing power as a universal quantum computer. A universal quantum computer can accurately and efficiently simulate the universe. The results of measurements made in the universe are indistinguishable from the results of measurement processes in a quantum computer. We can now give a precise answer to the question of whether the universe is a quantum computer in the technical sense. The answer is Yes. The universe is a quantum computer.

What is the universe computing? Everything we see and everything we don't see is a manifestation of the universe's quantum computation. We won't know exactly how the universe performs its most minute computations until we obtain a complete theory of fundamental physics, but even without knowing the full details, we can see that the quantum-computational power of the universe provides a direct explanation for its intricacy, diversity, and complexity.

Computation and Complexity

The universe we see outside our windows is amazingly complex, full of form and transformation. Yet the laws of physics are simple, as far as we can tell. What is it about these simple laws that allows for such complex phenomena?

To answer this question, first let's consider an old, and incorrect, theory of why the universe is complex. In the latter part of the nineteenth century, three physicists—James Clerk Maxwell, Ludwig Boltzmann, and Josiah Willard Gibbs—discovered that the thermodynamic quantity known as entropy was, as we've noted, a form of information: namely, information that isn't known. Inspired by concepts of information, Boltzmann proposed an explanation for the order and diversity of the universe. Suppose, said Boltzmann, that the information that defines the universe resulted from a completely random process, as if each bit were determined by the toss of a coin.

This explanation of the order and diversity of the universe is equivalent to a well-known scenario, apparently proposed by the French mathematician Emile Borel at the beginning of the twentieth century. Borel imagined a million monkeys (*singes dactylographes*) typing at typewriters for ten hours a day. Borel pointed out that over the course of a single year, the scripts the monkeys produced could conceivably contain all of the texts shelved in the world's richest libraries. (He then went on to dismiss the probability of this happening as infinitesimally small.)

Borel's image of monkeys pounding on the keys was subsequently appropriated by the British astronomer and mathematician Arthur Eddington, who settled on reproducing all the books in the British Museum. Eddington's version was taken up by Sir James Jeans, who erroneously ascribed it to Thomas Huxley. In Huxley's 1860 debate with Bishop Wilberforce regarding Darwin's *Origin of Species*, he did indeed mention monkeys. Wilberforce asked whether Huxley was descended from a monkey on his grandfather's or his grandmother's side, to which Huxley responded that he would rather be descended from a monkey than from a man of great intelligence who used his gifts in the service of falsehood. But none of the contemporary reports of the debate mention monkeys plugging away at typewriters (which had barely been invented at the time).

By the mid-twentieth century, the idea of monkeys inadvertently reproducing the world's literature had made it into the pages of *The New Yorker*, in Russell Maloney's short story "Inflexible Logic." Typing monkeys began to proliferate in the stories of Isaac Asimov, Douglas Adams, and others. A typical monkey-typing story begins with a researcher assembling a simian team and teaching them to hit the typewriter keys. One monkey inserts a fresh sheet of paper and begins to type: "Hamlet. Act I, Scene I. . . ."

It is certainly *possible* for one or another of these monkeys to type *Hamlet*. It is *possible,* as well, that the information defining the universe was created by similarly random processes. After all, if we identify heads with 1 and tails with 0, tossing a coin repeatedly will *eventually* produce any desired string of bits of a finite length, including a bit string that describes the universe as a whole.

The explicit argument against the creation of long texts by completely random processes dates back more than two thousand years, to Cicero. In his *De natura deorum* ("On the Nature of the Gods"), Balbus the Stoic presents the following argument against the atomists (such as Democritus), who have argued that the order of nature arose out of the random collision of atoms: "I can't but marvel that there could be anyone who can persuade themselves that solid atoms moving under the force of gravity could construct this elaborate and beautiful world out of their chance collisions. If they believe this could have happened, then I don't understand why they shouldn't also think that if innumerable copies of the twenty-one letters of the alphabet, made of gold or what have you, were shaken together and thrown out on the ground they could spell out the whole text of the *Annals* of Ennius. I doubt whether chance would succeed in spelling out a single verse!"

Let's rephrase Balbus's argument in terms of monkeys. Although the universe *could* have been created entirely by random flips of a coin, it is highly unlikely, given the finite age and extent of the universe. In fact, the chance of an ordered universe like ours arising out of random flips of a coin is so small as to be effectively zero. To see just how small, consider the monkeys once again. There are about fifty keys on a standard typewriter keyboard. Even ignoring capitalization, the chance of a monkey typing "h" is one in fifty. The probability of typing "ha" is one-fiftieth of one in fifty, or 1 in 2,500. The probability of typing "ham" is one in fifty times fifty times fifty, or 1 in 125,000. The probability of a monkey typing out a phrase with twenty-two characters is one divided by fifty raised to the twenty-second power, or about 10^{-38}. It would take a billion billion monkeys, each typing ten characters per second, for each of the roughly billion billion seconds since the universe began, just to have one of them type out "hamlet. act i, scene i."

By way of practical example, a popular Web site, http://user.tninet .se/~ecf599g/aardasnails/java/Monkey/webpages/, enlists your computer as a "monkey" in an attempt to reproduce passages from Shakespeare at random. The record as of this writing is the first twenty-four letters of *Henry IV, Part 2,* typed after 2,737,850 million billion billion billion monkey years.

The combination of very small probabilities together with the finite age and extent of the visible universe makes the completely random generation of order extremely unlikely. If the universe were infinite in age or extent, then sometime or somewhere every possible pattern, text, and string of bits would be generated. Even in an infinite universe, however, Boltzmann's argument fails. If the order we see were generated completely at random, then whenever we obtained new bits of information, they too would be highly likely to be random. But this is not the case: new bits revealed by observation are rarely wholly random. If you question this statement, just go to the window and look out, or pick up an apple and bite into it. Either action will reveal new, but non-random bits.

Here's another example: In astronomy, new galaxies and other cosmic structures, such as quasars, are constantly swimming into view.* If the argument for complete randomness were true, then as new objects swam into view they would reveal completely random arrangements of matter—a sort of cosmic slush—rather than the quasars and ordered, if mysterious, objects that we do, in fact, see.

In short, Boltzmann's explanation of order is not impossible. But it is hugely improbable.

Just for the fun of it, let's see how much of *Hamlet* could have been

---

*Note that the discovery of new cosmic objects does not come about simply because telescopes are getting better and better. As time goes on, the distance we can see increases, at the rate of about one light-year per year—so the number of objects we can see increases. In cosmology this phenomenon is known as the "expansion of our horizon," which is getting farther and farther away second by second.

generated by random processes since the universe began. The universe is full of photons—particles of light left over from the Big Bang. There are about 10^{90} photons, and each photon registers a few random bits. If we interpret those bits as characters in English, then somewhere out there is a bunch of photons that reads "*Hamlet,* Act I, Scene I. Enter Barnardo and Francisco." Even if we imagine that every elementary particle is a monkey that has been typing at the maximum rate allowed by the laws of physics since the universe began, the longest initial piece of *Hamlet* that could have been generated by random typing is "*Hamlet,* Act I, Scene I. Enter Barnardo and Francisco. Barnardo: Who's there?"

Just to create the first few lines of *Hamlet* by a fully random process such as monkeys typing would take the entire computational resources of the universe. To create anything more complicated by a random process would require greater computational resources than the universe possesses.

Boltzmann was wrong: the universe is not completely random. However, this does not mean that Cicero's Balbus was right. He was wrong, too. The existence of complex and intricate patterns does not require that these patterns be produced by a complex and intricate machine or intelligence. I'll say it again: Computers are simple machines. They operate by performing a small set of almost trivial operations, over and over again. But despite their simplicity, they can be programmed to produce patterns of any desired complexity, and the programs that produce these patterns need not possess any apparent order themselves: they *can* be random sequences of bits. The generation of random bits does play a key role in the establishment of order in the universe, just not as directly as Boltzmann imagined.

The universe contains random bits whose origins can be traced back to quantum fluctuations in the wake of the Big Bang. We have seen how these random bits can serve as "seeds" of future detail ranging from the positions of galaxies to the locations of mutations in DNA. These random bits, introduced by quantum mechanics, in effect *programmed* the later behavior of the universe.

Back to the monkeys. This time, instead of having the monkeys type random sequences of characters into *typewriters,* let's have them type random sequences into a *computer.* (The image of monkeys typing away at computers is ubiquitous, at least in cyberspace. I first heard about it from Charles Bennett and Gregory Chaitin of IBM in the 1980s.) For example, let's say we sit a monkey down at a PC, and tell the computer that the typescript is a program in a computer language, such as Java. The computer then interprets the monkey's random output not as a text, but as a computer program—that is, as a sequence of instructions in a particular computer language.

What happens when the computer tries to execute this random program? Most of the time, it will become confused and stop, issuing an error message. Garbage in, garbage out. But *some* short computer programs—and thus, programs with a relatively high probability of being randomly generated—actually have interesting outputs. For example, a few lines of code will make the computer start outputting all the digits of pi (π). Another short program will make the computer produce intricate fractals. Another short program will cause it to simulate the standard model of elementary particles. Another will make it simulate the early moments of the Big Bang. Yet another will allow the computer to simulate chemistry. And still another will start the computer off toward proving all possible mathematical theorems.

Why do computers generate interesting results from short programs? A computer can be thought of as a device that generates patterns: Any conceivable pattern that can be described in language can be generated by a computer. The key difference between monkeys typing into typewriters and monkeys typing into computers is that in the latter case the random bits they generate are interpreted as instructions.

As we've illustrated, almost nothing a monkey types exhibits a pattern in and of itself. When a monkey types random strings into a typewriter, all the typewriter does is faithfully reproduce those patternless strings. But when the monkey types into a computer, the computer interprets those patternless strings as instructions and uses them as a basis for constructing patterns.

Quantum mechanics supplies the universe with "monkeys" in the form of random quantum fluctuations, such as those that seeded the locations of galaxies. The computer into which they type is the universe itself. From a simple initial state, obeying simple physical laws, the universe has systematically processed and amplified the bits of information embodied in those quantum fluctuations. The result of this information processing is the diverse, information-packed universe we see around us: programmed by quanta, physics gave rise first to chemistry and then to life; programmed by mutation and recombination, life gave rise to Shakespeare; programmed by experience and imagination, Shakespeare gave rise to *Hamlet*. You might say that the difference between a monkey at a typewriter and a monkey at a computer is all the difference in the world.

Part 2

A CLOSER LOOK

CHAPTER 4

Information and Physical Systems

Information Is Physical

By now you know that the central theme of this book is that all physical systems register and process information, and that by understanding how the universe computes, we can understand why it is complex. So, when did the realization that all physical systems register and process information—something previously thought of as nonphysical—come about? The scientific study of information and computation originated in the 1930s and underwent explosive growth in the last half of the twentieth century. But the realization that information is a fundamental physical quantity predated the scientific study of either information or computation. By the end of the nineteenth century, it had been well established that all physical systems register a definable quantity of information and that their dynamics transform and process that information. In particular, the physical quantity known as entropy came to be seen as a measure of information registered by the individual atoms that make up matter.

The great nineteenth-century statistical physicists James Clerk Maxwell in the United Kingdom, Ludwig Boltzmann in Austria, and Josiah Willard Gibbs in the United States derived the fundamental formulas of what would go on to be called "information theory," and used them to characterize the behavior of atoms. In particular, they applied these formulas in order to derive justification for the second law of thermodynamics.

As noted, the first law of thermodynamics is a statement about *energy:* energy is conserved when it is transformed from mechanical energy to heat. The second law of thermodynamics, however, is a statement about *information,* and about how it is processed at the microscopic scale. The law states that entropy (which is a measure of information) tends to increase. More precisely, it states that each physical system contains a certain number of bits of information—both invisible information (or entropy) and visible information—and that the physical dynamics that process and transform that information never decrease that total number of bits.

Although it is more than a century and a half old, the second law of thermodynamics remains a subject of scientific controversy. Almost no scientist doubts its truth, but many disagree as to *why* it is true. The computational nature of the universe can resolve at least part of this controversy. Properly understood, the second law of thermodynamics rises from the interplay between "visible information," the information we have access to about the state of matter, and "invisible information," the bits of entropy—no less physical—that are registered by the atoms forming that matter.

Origins of the Computational Model

My undergraduate curriculum at Harvard went by the name "General Education." In practice, this seemed to mean that if I could talk my way into a course, then it was part of my curriculum. Accordingly, with the blessing—or, at any rate, the signature—of my advisor, the Nobel laureate Sheldon Glashow, I designed my undergraduate physics curriculum around Robert Fitzgerald's courses on prosody and on Homer, Virgil, and Dante, supported by Leon Kirchner's course on chamber music performance and I. Bernard Cohen's graduate seminar on the Influences of the Physical Sciences on the Social. Glashow also insisted that I take some physics.

Two courses I took sent me down the path that would lead to the computational model of the universe. The first was Michael Tinkham's course in statistical mechanics, the remarkable synthesis of quantum mechanics (the physics of atoms and molecules) and thermodynamics (the study of heat and work). As a science, statistical mechanics began in the last years of the nineteenth century and has led to lasers, lightbulbs, and transistors, to name just a few of its consequences. The primary message of Tinkham's course was that the thermodynamic quantity called entropy—known as a measure of the heat energy that can't be turned into mechanical energy in a closed thermodynamic system—can also be understood as a measure of information.

Entropy (from the Greek for "in turning") was first defined by Rudolf Clausius in 1865 as a mysterious thermodynamic quantity that limits the power of steam engines. Heat has lots of entropy. Engines that run off of heat, like steam engines, have to do something with that entropy; typically they get rid of it in the form of exhaust. They can't turn all of the energy in heat into useful work. Entropy, said Clausius, tends to increase.

At the end of the nineteenth century, the founders of statistical mechanics—Maxwell, Boltzmann, and Gibbs—realized that entropy was also a form of information: entropy is a measure of the number of bits of *unavailable* information registered by the atoms and molecules that make up the world. The second law of thermodynamics comes about, then, by combining this notion with the fact that the laws of physics preserve information, as we will soon discuss. Nature does not destroy bits.

But surely it takes an infinite number of bits of entropy to specify the positions and velocities of even a single atom exactly, my class objected. Not so, said Tinkham. The laws of quantum mechanics, which govern the microscopic behavior of physical systems, ensure that atoms and molecules register a finite amount of information.

Hot bits! This was great stuff, even if I didn't fully understand it. All physical systems can be characterized in terms of information, and Maxwell, Boltzmann, and Gibbs had figured this out fifty years before the word "bit" was even invented! But what about this quantum mechanics? Clearly I needed to know more. So I took Norman Ramsey's introductory course on quantum mechanics. Ramsey is one of the world's most expert quantum-mechanical masseurs. He developed many of the techniques used today to convince atoms and molecules to give up their energy and their secrets, techniques for which he went on to win a Nobel Prize.

But what was plain to Ramsey about quantum mechanics remained opaque to me. How could it be, for example, that an electron can be in two places at once? Ramsey assured us through detailed experimental data that not only was an electron *allowed* to be in many places at the same time, it was in fact *required* to be there (and there, and there, and there). Perhaps the early hour in the lecture hall, lit only by the glow of a transparency projector, had induced a trancelike state—but I didn't get it. I would not awaken from this particular trance until years later, when I was working for Ramsey at the Institut Laue-Langevin in Grenoble, France, on his experiment to measure separation of electric charge inside the neutron.

The neutron and its charged partner, the proton, are the particles that make up the nuclei of atoms. Neutrons and protons are in turn made up of electrically charged particles called quarks. The separation of electric charge that Ramsey wanted to measure corresponded to a distance of one billion billion billionth of a meter between the quarks within the neutron, a distance smaller, relative to the size of the neutron, than the size of the neutron, relative to us. The experiment involved taking neutrons from a nuclear reactor, cooling them down until they were moving at walking pace (in the final cooling stage, the neutrons were made to run uphill until they were exhausted and almost came to a halt), subjecting them to electric and magnetic fields, and then "massaging" them into a state in which they would reveal their secrets.

As you may guess, you have to massage a neutron very sensitively for it to reveal anything at all. Everything has to go right in such an experiment, or nothing happens. Our neutrons were fickle, and no matter how many times we polished their electrodes and pumped out their vacuum, they refused to talk to us. In this slack time, Ramsey assigned to me a simple calculation based on the neutron spinning both clockwise and counterclockwise at the same time, all the while conversing with the particles of light around it. Perhaps it was because this was the first time I had ever been asked to do a calculation for a real experiment, or perhaps it was because Ramsey snapped his fingers, but I awoke from my trance. Neutrons, I saw, *had* to spin clockwise and counterclockwise at the same time. They had no choice: it was in their nature. The language that neutrons spoke was not the ordinary language of yes *or* no, it was yes *and* no at once. If I wanted to talk to neutrons and have them talk back, I had to listen when they said yes and no at the same time. If this sounds confusing, it is. But I had finally learned my first words in the quantum language of love.* You will learn to say a few words in this language yourself in the next chapter.

Michael Tinkham's course on statistical mechanics taught me that physical objects might be thought of as being made of information. Ramsey's course on quantum mechanics taught me how the laws of physics governed the way in which that information was represented and processed. Most of the scientific work I have done since I took those courses has revolved around the interplay between physics and information. The computational nature of the universe itself arises from this interplay.

The Atomic Hypothesis

The mathematical theory of information was developed in the middle of the twentieth century by Harry Nyquist, Claude Shannon, Norbert

---

*Ramsey also gave me a good lesson in the language more usually called the language of love. I happened to be in his office when two members of the Academie Française came to call. "Why, Professeur Ramsey," they inquired, "is French not the international language of Science?" Ramsey immediately answered them in his fluent French, with a thick midwestern accent. Horrified, they dropped the subject. In fact, the French Academy of Sciences caused the adoption of English as the international language of science in the seventeenth century by being the first national academy to abandon the previous international language, Latin, and publish their proceedings in their own language. The English and the Germans followed suit. The rest is just an accident of history.

Wiener, and others. These researchers used mathematical arguments to derive formulas for the number of bits of information that could reliably be sent down communication channels such as telephone lines. When Shannon showed his new formula for information to the mathematician John von Neumann and asked him what the quantity he had just defined should be called, von Neumann is said to have replied, "*H*."

"Why *H?*" asked Shannon.

"Because that's what Boltzmann called it," said von Neumann. The basic formulas for information theory had already been derived by Maxwell, Boltzmann, and Gibbs.

To understand what information has to do with atoms, look at the origins of the atomic hypothesis. The ancient Greeks postulated that all matter was made of atoms (the Greek *atomos* meant "unsplittable"). The atomic hypothesis was based on an aesthetic notion: distaste for the infinite. The ancients simply did not want to believe that you could keep subdividing matter into ever smaller pieces. Isaac Newton's and Gottfried Wilhelm Leibniz's invention of calculus in the seventeenth century, however, provided mathematical methods for dealing with the infinitely small, and early attempts to describe solids, liquids, and gases mathematically modeled them as *continuous* substances that could be subdivided an infinite number of times. The power and elegance of calculus, together with the lack of direct evidence for the existence of atoms, made for scientific theories based on the continuum. But by the second half of the nineteenth century, observational evidence had begun to indicate that, as proposed by the atomic hypothesis, matter might indeed be made up of very small, discrete chunks, rather than being continuous.

For example, if you look through a microscope at tiny dust particles suspended in a liquid, you see them doing a jiggly dance called Brownian motion. This jiggling is caused by the dust particles being bombarded from all directions by molecules of the liquid in which the dust is suspended. When, by chance, more molecules hit the dust from the left than from the right, the dust particle recoils to the right. When more molecules hit the dust from the top, it recoils to the bottom. At the beginning of the twentieth century, Einstein provided an elegant quantitative theory of Brownian motion, showing that the observed motion was consistent with bombardment of the suspended dust particles by much smaller particles, of a particular size and mass. The atomic hypothesis was back.

Prior to Einstein's work, the atomic hypothesis had been used to provide a rigorous basis for the behavior of heat and energy. Heat had long been known to be a form of energy. In the eighteenth century, Sir Benjamin Thompson, Count Rumford performed a famous demonstration in which he immersed in water a cannon that was being bored by a horse-powered drill. The horses went round and round in a circle, turning a drill that removed metal to form the cannon's bore. The water eventually boiled, showing the conversion of horsepower into heat. The quantitative trade-off between mechanical energy and heat was even more firmly established by the mid-nineteenth century and enshrined as the first law of thermodynamics: energy is conserved when mechanical energy is converted to heat. Unlike mechanical energy, however, energy in the form of heat seemed to possess the mysterious property called entropy, which prevented some of the heat from being transformed into useful work. Like energy, entropy could be quantified experimentally: Whenever mechanical energy was turned into heat, an amount of entropy equal to the energy divided by the temperature was created. When heat was turned into mechanical energy, as in one of Watt's steam engines, the amount of entropy in the cooler steam of the engine's exhaust was discovered to be either greater than or equal to the amount of entropy in the hot steam driving the engine. Entropy, whatever it was, never decreased.

Just what is this entropy stuff, anyway? The atomic hypothesis provides an answer. Heat is a form of energy, and entropy is associated with heat. If things are made out of atoms, then there is a simple explanation of heat: heat is just the energy in the jiggling of atoms. Entropy, then, has a simple interpretation, too: To describe the motion of atoms requires a large number of bits of information. The quantity called entropy is proportional to the number of bits required to describe the way atoms are jiggling.

For the scientists of the nineteenth century, it wasn't a stretch of the imagination to think of heat as the energy in jiggling atoms. After all, since the work of Galileo and Newton hundreds of years earlier, it had been known that everything that moves has energy—called kinetic energy, from the Greek *kinesis,* "motion"—associated with that motion. The faster a thing moves, the more kinetic energy it has. When mechanical energy is converted to heat, as when horses bore a cannon and heat water, the mechanical energy generated by the horses is converted into the kinetic energy of the molecules of water. Similarly, when hot gas moves a piston in a steam engine, it is because the water molecules that form the steam are exerting pressure on the piston head as they bounce off it. When mechanical energy is converted into the kinetic energy of atoms and molecules, and vice versa, the first law of thermodynamics guarantees that the overall energy is conserved. It was not so natural, though, for nineteenth-century scientists to conceive of *entropy as information.* Nowadays, in the midst of the information-processing revolution, it is no longer surprising that information should be just as fundamental a quantity as energy, but at the end of the nineteenth century, it was not even clear that information was a quantity at all.

In the middle of the nineteenth century, James Clerk Maxwell developed a detailed theory of heat in terms of the motion of atoms. He figured out how fast the atoms were moving as a function of temperature: the kinetic energy of an atom is proportional to its temperature. The hotter something is, the faster its atoms are jiggling around.

This jiggling is also associated with entropy: the faster the atoms jiggle, the more information is required to describe their jiggling, and thus, the more entropy they possess. Temperature is a measure of the trade-off between information and energy: atoms at a high temperature require more energy to register a bit of information, and atoms at a low temperature require less energy to register a bit. Temperature is energy per bit. When energy in the form of heat flows from a hot thing to a cold thing, entropy increases: the same amount of energy registers less information when it's hot than when it's cold. The state of maximum entropy is obtained when everything is at the same temperature.

Maxwell realized that if one could gain information about the microscopic behavior of the atoms in a gas, one could reduce its entropy: entropy was somehow associated with information. In a famous letter, "On the Decrease of Entropy by Intelligent Beings," Maxwell imagined a tiny intelligent being, or "demon," who could make heat flow from cold bodies to hot, thereby apparently violating the second law of thermodynamics.

Figure 5. Maxwell's Demon

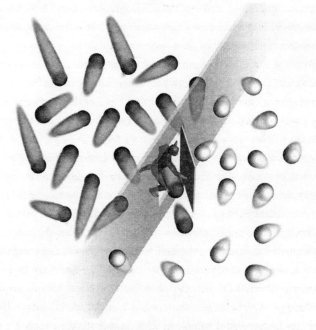

Maxwell's demon is an imaginary being that shunts fast, hot molecules to one side of a container and slow, cold molecules to the other side in apparent violation of the second law of thermodynamics.

Suppose a container of helium gas is divided in half by a partition. In the partition is a small door, just big enough for a few gas atoms to pass through at a time. The demon monitors the atoms in the neighborhood of the door and opens the door whenever the atoms that are approaching the door from the colder side are jiggling faster (thus are hotter) than the atoms approaching it from the hotter side. Each time the demon opens the door, hotter atoms move to the hot side, and cooler atoms move to the cold side. As the demon sorts more and more fast-jiggling atoms into the hot side and more and more slow-jiggling atoms into the cold side, the hot side of the gas grows hotter and the cold side grows colder. This demon-mediated flow of heat from the cold side to the hot side apparently violates the second law of thermodynamics, which implies that heat flows from hot to cold but not from cold to hot. *It is the demon's ability to get information about the atoms that allows him to accomplish this apparent violation of physical law.*

As we will see, the demon cannot actually violate the second law, which states that the total entropy/information of the gas and the demon combined can't decrease. The second law of thermodynamics remains intact. But still, a clear connection between information and entropy is illustrated by his demonic feat.

As the nineteenth century wound on, Boltzmann, Gibbs, and the German physicist Max Planck refined their formulas describing the energy and entropy of systems made up of atoms. In particular, they discovered that the entropy of a system was proportional to the number of bits required to describe the microscopic state of the atoms. This result was so useful in describing the trade-offs between heat and energy that the formula that encompasses it is inscribed on Boltzmann's tomb. Entropy is traditionally written S, and the number of different possible microscopic states (or "complexions," as Planck called them) is W. W could be the number of complexions of an individual atom or of a system made of atoms. The epitaph on Boltzmann's tomb reads "$S = k \log W$," which is just a fancy way of saying that the entropy of something is proportional to the number of bits registered by its microscopic state.

Another way of saying the same thing is that the entropy is proportional to the length, in bits, of the number of the microscopic states. In this formula, k is known as Boltzmann's constant.

Paul and Tatyana Ehrenfest, who made many original contributions to the study of entropy, point out that this formula was actually first written down by Planck, and they therefore refer to the constant that we call Boltzmann's constant as "Planck's constant." But, as we'll see when we look at quantum mechanics, Planck already had a rather important constant named after him, so in order to avoid confusion, as well as to honor Boltzmann, k was renamed for him. Boltzmann was famously moody, and died by his own hand shortly after a visit to the United States in 1906. One wonders what he would have made of another man's formula being inscribed on his tomb.

Maxwell, Boltzmann, Gibbs, and Planck discovered that entropy is proportional to the number of bits of information registered by the microscopic motions of atoms. Of course, these nineteenth-century scientists did not think of their discovery as being primarily about information. At the time, however, entropy was not measured in bits, so they considered their discovery to be the proper expression for thermodynamic entropy—the quantity that limits the efficiency of heat engines. They were correct. And since entropy was not measured in bits at that time, the measurement had to be multiplied by Boltzmann's constant to relate entropy as measured in terms of information to ordinary thermodynamic entropy. But whether they knew it or not, the pioneers of statistical mechanics discovered the formula for information fifty years before the mathematical theory of information was in place.

Just how does a physical system, such as a gas, register information? Consider a child's balloon filled with helium. The helium atoms in the balloon are zipping around from place to place inside the balloon, bouncing off each other and off the walls of the balloon. Each helium atom registers information: the amount of information required to describe where it is (position) and where and how fast it is going (velocity). In order to measure the amount of information an atom registers, you must define the smallest scale (degree of precision) to which an atom's position and velocity can be described. Then the number of bits a given atom registers is equal to the number of bits required to specify its position and velocity to a precision given by that smallest scale. Later, we'll see that quantum mechanics defines the smallest scale to which position and velocity can be specified. Based on that scale, we can determine that each atom in the balloon registers about 20 bits. The amount of information registered by all the helium atoms in the balloon, then, is the number of atoms—6×10^{23}—times the number of bits per atom: a million billion billion (10^{25}) bits of information.

That's a lot of information. This book contains a few million bits of information. The millions of books in the Library of Congress contain some million millions of bits. All the computers in the world at present contain some billion billion bits. All the bits of information generated in written or electronic form by the human species as a whole still falls short of the amount of information registered by the atoms of helium in a balloon.

Mind you, the bits of information registered by the helium atoms in a balloon wouldn't make a very good read. Like the texts written by a monkey on a typewriter, the bits registered by the atoms would have a very high probability of looking like a bunch of random gibberish. Even if the positions and velocities of the helium atoms happened to spell out the whole of *Hamlet* at one instant in time (and we already know how unlikely that is), a second later their bits would bounce back into randomness.

Landauer's Principle

The second law of thermodynamics holds that the total amount of information never decreases. In our balloon, for example, the second law implies that the number of bits of information registered by the helium atoms doesn't get smaller if the balloon is left alone. If you cool the balloon, or squeeze it, or pop it, the number of bits registered by the atoms in the balloon can indeed decrease—but only at the expense of increasing the number of bits registered by atoms outside the balloon.

Information can be created but it can't be destroyed. Consider flipping a bit. Flipping a bit transforms information: 0 goes to 1 and vice versa. It also preserves information: if you knew that the bit was 0 before the flip, then you know that it is 1 after the flip.

By contrast, erasure is a process that destroys information. During erasure, a bit that is initially 0 stays 0, and a bit that is initially 1 goes to 0. Erasure destroys the information in the bit. But the laws of physics do not allow processes that do nothing but erase a bit. Any process that erases a bit in one place must transfer that same amount of information somewhere else. This is known as Landauer's principle, after Rolf Landauer, the pioneer of the physics of information, who discovered it in the early 1960s.

To see Landauer's principle in action, look at how bits are erased in computers. As noted in chapter 2, in a contemporary electronic computer a bit is stored on a capacitor. A capacitor is a bucket for electrons. When you charge up the capacitor, you put electrons in the bucket; when you discharge it, you dump the electrons out of the bucket. In a computer, an uncharged capacitor registers a 0 and a charged capacitor registers a 1.

To erase a bit in an electronic computer, just empty the bucket: close a switch and let the electrons on the capacitor flow out. When the capacitor has been discharged, the bucket is empty: the bit has been restored to 0. But the microscopic state of the electrons "remembers" whether the capacitor was charged or not; that is, as they flow out of the capacitor, they heat up. This temperature change remains as evidence of the initial state of the capacitor. Its bit of information has been transferred to the microscopic motion of the electrons.

Another way to erase a bit is to swap it with another bit that reads 0. Swapping information between bits preserves information; to get back the original values of the bits, just swap them again. Before the swap, the first bit could read either 0 or 1; it has a bit's worth of entropy. The second bit reads 0; it has no entropy. After the swap, the first bit reads 0; it has been restored to 0, or erased. The second bit reads 0 or 1; it has a bit's worth of entropy—the same entropy that the first bit had before the swap. Swapping moves information and entropy from one place to another, but the overall amount of information remains constant. Swapping can be used to erase a bit in one register while retaining a copy of the bit in another register. To return to the example of a computer's capacitor, discharging, or erasing its bit, essentially "swaps" the information registered by the capacitor with the information registered by the electrons.

The laws of physics preserve information as it is transformed. In mathematical parlance, the dynamical laws of physics of a closed physical system are one-to-one: *Each input state goes to one and only one output state, and each output state can have come from one and only one input state.* Thus, you can work backwards: if you know the physical state of a system now, then in principle you can follow its physical dynamics to determine the state of the system at earlier or later times.

So, for example, if you knew the exact state of the gas of helium atoms in the balloon at one point in time and were able to follow the detailed dynamics of the atoms bouncing off one another and off the inner wall of the balloon, then—because each state evolves dynamically into a completely determined state—you would know the state of the helium atoms at later times as well. Conversely, because each state evolves from a completely determined state, knowledge of the state now, together with the ability to follow the detailed dynamics, determines the state at previous times. In the case of flipping a bit, if you know what the state of the bit is before the flip, then you still know it afterward. Physical dynamics preserve information.

This preservation is what prevents heat engines like steam engines or automobile engines from extracting all the energy from heat. There's lots of energy in hot gas, but lots of bits as well. The temperature of the gas is proportional to the average energy per bit. Hot gas has more energy per bit; cold gas has less energy per bit. When the heat energy of the gas is extracted—for example, by having the gas push against a piston—the bits are left behind. The moving piston turns heat energy into mechanical energy, the energy per atom (and hence per bit), decreases, and the expanding gas cools down. As long as the temperature of the gas does not go to absolute zero, each atom (and hence each of its bits) still requires some energy, so that amount of energy must remain in the gas, rather than becoming mechanical energy. Since some energy must remain behind, not all of the energy can be extracted in the form of work.

Over the centuries, many ingenious inventors have proposed machines that promise to extract more energy than should be possible by this rationale. They've attempted to defy the second law of thermodynamics. Such a machine is traditionally called a *perpetuum mobile,* a perpetual motion machine.* As you might guess, these machines do not work because they fail to provide for the extra information. You would think that after centuries of failed effort, people would have given up on perpetual motion machines. Over the last fifteen years, however, I have been the referee for a number of ingenious proposals that try to extract more energy from physical systems than is allowed by the second law. They all fail. With practice, one can look at the most complicated of these schemes and see just where the inventor has swept the information under the rug.

The Spread of Ignorance

The laws of physics preserve information. The number of bits registered by a system (such as the helium-filled balloon) does not decrease. This information-preserving feature puts limits on the efficiency of heat engines and is responsible for the second law of thermodynamics. But now there's a problem. According to the laws of physics, total informa-

*Technically, a device that tries to extract work from heat without any exhaust to get rid of the information is called a perpetual motion machine of the second kind; perpetual motion machines of the first kind try to run forever without ever turning energy into heat or vice versa.

tion cannot *increase* either. In fact, the laws of physics say that in the absence of interaction with another system, the amount of information in a system stays the same. So, how can entropy—a form of information—increase without increasing the overall information content of a physical system? How does known information become unknown?

As originally conceived, entropy is a quantity that measures how *useful* energy is. Energy with a small amount of entropy is useful (free) energy; energy with lots of entropy is useless. It is perhaps easier to conceive of an increase in entropy in these terms: energy degrading from useful to useless forms. Hot baths grow cold. Cars run out of gas. Milk goes sour. So how can we think of this process in terms of information? The answer lies in a fundamental fact of nature that I call "the spread of ignorance." Unknown bits infect known bits.

We have seen that entropy is information about the microscopic jigglings of atoms—jigglings far too small for us to see even with the most powerful microscope. Each of the helium atoms in our balloon registers twenty bits. But unless we know where an individual atom is in the balloon and how fast it is going (up to the accuracy allowed by quantum mechanics), we have no idea what those bits are. In other words, entropy, which is just invisible information, is also a measure of ignorance.

We do have some information about the atoms in the balloon. For example, we can measure the macroscopic state of the balloon: its size, its temperature, the pressure the atoms exert on its walls. Typically, we possess only a few hundreds of bits of macroscopic information about a physical system such as a balloon. For any system, we can make a distinction between bits whose values (0 or 1) we know and those whose values we don't know. Bits with values of which we are ignorant constitute the entropy of the system: a bit of entropy is a bit of ignorance.

Note that the division between known and unknown information is to some degree subjective. Different people know different things. For example, suppose you send me a brief e-mail containing 100 bits of information. You know what those bits are, because you sent them. To you, the information in the e-mail is known. Before I open the e-mail, I don't know what those bits are: they are still invisible to me. At this point, I would count its 100 bits as a form of entropy. So different observers can assign different values to the entropy of a system. Remember Maxwell's demon? Since it is monitoring the microscopic state of the gas, it has more information than an observer who simply knows the gas's temperature and pressure. Accordingly, the demon assigns the gas a lower entropy than the macroscopic observer does. For the purpose of the second law of thermodynamics, the important quantity is the *total* amount of information in a physical system. The total amount of information, known and unknown, in a physical system does not depend on how it is observed.

Suppose an unknown bit of information interacts with a known bit of information. After the interaction, the first bit is still unknown, but now the second bit is unknown, too. The unknown bit has infected the known bit, spreading the lack of knowledge, and increasing the entropy of the system. We can use the ideas of computation developed earlier to clarify this picture of the infectious nature of ignorance.

Consider two bits. The first bit is unknown: it could read either 0 or 1. The second bit is known to be, say, 0. Thus, the two bits together are either in the state 00 or 10. Now apply the following simple logic operation to the bits. Flip the second bit if and only if the first bit is 1. This operation is called a controlled-NOT op, because it performs a bit-flip (or NOT operation) on the second bit, an operation *controlled* by the state of the first bit (which in this case is unknown). If the first bit is 1, then the controlled-NOT op will flip the second bit from 0 to 1. If the first bit is 0, the controlled-NOT op will leave the second bit a 0. After the controlled-NOT op, the two bits taken together will either be in the state 00 or in the state 11. The two bits are now correlated—that is, they have the same value. If we look at the first bit, we know the value of the second bit, and vice versa.

After the op, the first bit is still unknown: it could still be in either the state 0 or the state 1. But look at the second bit. Now it, too, could be in either the state 0 or the state 1. The second bit, which was known to be 0 before the controlled-NOT operation, is now unknown, too. The controlled-NOT operation has caused the unknown information in the first bit to infect the second bit—the ignorance has spread! (The spread of ignorance is reversible. To get the original state of the two bits back, just perform the controlled-NOT twice. The controlled-NOT operation is its own inverse: applying it twice is like doing nothing at all.)

The spread of ignorance increases the entropy of the individual bits in a system. The entropy of the first bit remains at one bit, but the entropy of the second bit has increased. But the entropy of the bits *taken together* remains constant. Before the controlled-NOT, the two bits could have been in one of two states, 00 or 10. There is one bit of entropy, all in the first bit. After the controlled-NOT operation, the two bits could still be in one of two states, 00 or 11. There is still only one bit of entropy, but now it is spread out between the two bits.

The spread of ignorance is reflected in the increase of a quantity called "mutual information." Each bit, on its own, has one bit's worth of entropy, but the two bits taken together also have only one bit's worth of entropy. The mutual information is equal to the sum of the entropies taken separately, minus the entropy of the two bits taken together. In other words, the two bits have exactly one bit of mutual information. Whatever information they have is held in common.

Atomic Ignorance

The infectious nature of information applies to colliding atoms as well as to bits in a computation. The argument that the entropy of the individual atoms in a gas tends to increase was originally put forward by Ludwig Boltzmann in the 1880s. Boltzmann defined the quantity he called H as the degree to which we know the position and velocity of any given atom in a gas.

Boltzmann's quantity H is in fact the entropy of an individual atom, multiplied by minus one. He showed that when the positions and velocities of the atoms are uncorrelated—that is, independent of each other—collisions between them will decrease H and increase the entropy of the individual atoms. Subsequent collisions, he argued, would continue to increase that entropy. He concluded that his H-theorem justified the second law of thermodynamics by supplying a mathematical proof that entropy must increase.

The problem with Boltzmann's H-theorem is that it is not, strictly speaking, true of actual atoms in a gas. Boltzmann was quite correct that collisions between initially uncorrelated atoms will increase the atoms' individual entropies. These entropies increase because of the infectious nature of information. When two atoms collide, any uncertainty about the position and velocity of the first atom tends to infect the second atom, rendering its position and velocity more uncertain and thus increasing its entropy. This entropy increase of the second atom is analogous to the entropy increase of the second bit described above, when that bit was subjected to a controlled-NOT operation with an unknown bit as the controller.

The flaw in the H-theorem lies with subsequent collisions. Once two atoms have collided and exchanged information, subsequent collisions can *decrease* the entropy of the individual atoms. To understand how interaction between two atoms that have previously collided can decrease their entropy, return to the two bits above. The first time the controlled-NOT operation is applied, the entropy of the control bit infects the second bit, increasing its entropy by one bit. But if the controlled-NOT operation is applied again, the second bit is restored to its initial, known state, decreasing its entropy by one bit.

In principle, a similar inverse operation, resulting in a similar entropy decrease, could be engineered for the atoms. When Boltzmann presented his H-theorem as a proof of the second law of thermodynamics, his colleague Joseph Loschmidt pointed out that the H-theorem couldn't always be true, since reversing the velocities of the atoms would undo the collision and decrease their entropies. (The hypothetical being that could reverse the velocities of the atoms is known as Loschmidt's demon. Back in those days, everyone had demons.) Confronted with this (correct) argument, Boltzmann resorted to sarcasm: "Go ahead," he said, "reverse them."

Boltzmann's original argument for his H-theorem relied on an assumption about the nature of atomic collisions called "the assumption of molecular chaos." Even though the positions and velocities of two atoms might be correlated before their collision, Boltzmann argued, repeated collisions between many atoms tend to dilute that correlation, so that two colliding atoms in a gas would in effect be uncorrelated at the moment of their collision. Right after they collide, the two atoms' positions and velocities are correlated. But as they go on to collide with other atoms, their correlations with each other tend to fade. Boltzmann argued that the next time they collide, they can be assumed to be uncorrelated: it is as if the two atoms had never collided before. If the assumption of molecular chaos is true, then the entropies of the individual atoms almost always increase. This increase can be undone in principle by reversing the process of collision, à la Loschmidt. But in practice such a reversal rarely happens.

The assumption of molecular chaos is a good one and is true for many complex systems, such as gases. It is not true for all physical systems, however. As will be seen, in many physical systems it is possible to reverse the interactions between the pieces of the system, thereby reversing the entropy increase of those pieces.

By and large, however, Boltzmann's assumption works well. Even after atoms have collided once, their further collisions tend to increase their individual entropies. Why does the assumption of molecular chaos work so well? In my M.Phil. thesis, "The Spread of Ignorance," and Ph.D. thesis, "Black Holes, Demons, and the Loss of Coherence," I provided an answer for this question by developing an approach for treating the second law of thermodynamics in terms of the spread of ignorance. This method shows that Boltzmann's H-theorem is "almost true" for "almost all" physical systems.

Snooker

Some background to my approach is in order. After graduating from Harvard, I went to Cambridge University on a Marshall Scholarship. These scholarships are given by the British government out of gratitude for the Marshall Plan, which helped rebuild Europe after World War II. (Gratitude goes only so far. On my first day in Cambridge I went to a pub called the Locomotive. The fellow sitting next to me at the bar had spiky green hair and was wearing a dog collar. When I mentioned to him that his government was paying for me, an American, to attend Cambridge, he ungratefully insisted that I leave the premises.) My first year at Cambridge was spent taking Part III Maths, a course in mathematics and physics one of whose goals is to identify promising scholars and weed out the rest. Students who obtain first-class degrees in Part III typically go on to do Ph.D.s. The very top students are known as wranglers (yippee-ti-yi-yo!). Maxwell had been a wrangler. As for the rest—well, let's just say that the reward for the bottom student at graduation used to be a four-foot-long wooden spoon.

To become a wrangler required ceaseless application to study. Many of my classmates were cocooned in the library throughout Part III. Their personalities would not emerge until after they graduated. I had read about student life at Cambridge in the novels of E. M. Forster and the poems of Wilfred Owen. Though I wished to avoid the spoon, I knew that if there was a branch of physics I wanted to study, it was the interplay between mechanics and fluid dynamics involved in rowing in a gentleman's eight or punting to Grantchester. After my morning lectures, I would go off to the pub by the river and down the lane from the Department of Applied Math and Theoretical Physics (or DAMTP, encouragingly pronounced "damped") to eat a Cornish pasty and drink a pint of Guinness. Then I would go either for a row on the river or to the graduate student lounge for a game of snooker.

Snooker is a game related to pool. It is played with cues and balls on a table much larger than a pool table. The special cue employed to reach for shots at the far end of the table could equally well be used for pole vaulting. Snooker shares with cricket, lawn bowling, and sheepherding the classic feature of televised British sports: it is played on a vast green expanse on which small objects (men, balls, sheep) are distributed. Snooker is also like pool in that its goal is to use a cue ball to knock colored balls into pockets, a procedure known as potting. But in snooker, unlike pool, one alternates between potting red balls and potting yellow, blue, pink, or black balls.

The secret to entropy increase is to be found in snooker. The collision of two snooker balls contains in two dimensions almost all the elements of the collision between two helium atoms in three. At the beginning of the game, the balls all start out in specified positions with zero velocity: their entropy is small. After a few shots, they are all over the table, in positions that depend sensitively on the past history of collisions between the balls and on slight variations in how they were struck by the cue. That uncertainty in how the cue ball is struck—a few bits of unknown information—infects all the balls with which the cue ball collides.

In the early part of the twentieth century, Emile Borel (he of the monkeys typing) suggested that entropy increase could be thought of as arising from the interactions between systems spreading information around. Starting from Borel's observations, my thesis work showed that interactions between pieces of a system, such as atoms in a gas—or snooker balls on a table—tend to increase the entropies of those pieces, even if they have interacted before. This result justifies Boltzmann's assumption of molecular chaos, because it implies that a collision between two atoms will almost always increase their entropies even if they have collided before. Eventually the entropies of the individual parts of a system such as a gas tend to rise to their maximum possible value.

Figure 6. Pool and the Second Law of Thermodynamics

6a. *The balls are in a low-entropy, triangular arrangement, but the cue ball is heading toward them.*

6b. *After the cue ball breaks up the array, the balls move off in a pattern whose entropy and randomness increase with every collision.*

When atoms bounce off each other, they exchange information and spread entropy. Any ignorance about the state of one atom spreads to the state of the other. The spread of ignorance is also familiar in snooker, where the same arguments apply. The cue ball imparts part of its velocity (that is, some of its bits) to the red ball. The red ball bounces off the pink ball, spreading some of its bits, including those it got from the cue ball, to the velocity of the pink. As more and more collisions take place, the number of bits of ignorance distributed among the balls increases, until the bits (and the balls) are spread all over the table. Bits are infectious.

A particularly interesting case of this bit-contamination process arises when some of the information about a system is macroscopic (that is, information we can access directly through observation and measurement) and the rest is microscopic, "invisible" information (or entropy). Over time, we would expect the microscopic, hidden information to infect the macroscopic, observable information. Eventually the information and entropy of all the system's bits tends to its maximum allowed value.

This infection of macroscopic bits by microscopic ones is a feature of chaos. Recall that a chaotic system is one whose dynamics tend to amplify small perturbations, so that microscopic information is pumped up to the macroscopic regime. In a chaotic system, the invisible information in the microscopic bits infects the macroscopic bits, causing the observable characteristics to wander in an uncertain fashion—like the butterfly's effect on the course of a hurricane.

The collision of snooker balls is also a chaotic process. Suppose you make a small error in striking the cue ball, so that its initial speed and direction is a bit off. That error is amplified when the cue ball strikes the red ball. The direction in which the red ball now moves has a greater error than the error in the initial speed and direction of the cue ball. The more collisions that take place, the more the initial error is magnified. If you planned to knock the red ball off the pink ball and knock that off the third to pot the third ball, you will probably have failed: by the third collision, the initial error has typically grown too great to afford any measure of control over the speed and direction of the third ball.

Ignorance spreads, individual entropies increase. In this picture of the second law of thermodynamics, entropy increase is like an epidemic. Bits of ignorance are like viruses that are copied and spread by interaction. The contagion continues until all parts of the system have been infected. At this point, the entropies of the parts taken individually are close to their maximum value.

The Spin-Echo Effect

When Joseph Loschmidt suggested that it might be possible to decrease the entropy of a gas by simultaneously reversing the velocities of all its atoms, Boltzmann taunted him. But as we'll now see, it is possible to realize Loschmidt's proposal in actual physical systems. In such systems, entropy can appear to decrease, in apparent (but not actual) violation of the second law of thermodynamics.

What happens if you reverse the motions of the components of a system? The interactions between the pieces of the system can undo themselves, decreasing their entropies. Loschmidt's original proposal—to reverse the velocities of the atoms in a gas—is impractical. But for some systems, when Boltzmann challenges you to "reverse them," you can.

A simple example of such a reversible dynamics is the controlled-NOT operation described earlier. In this straightforward logic operation, one bit is flipped if and only if a control bit reads 1. As noted, if the second bit is initially 0 and the control bit can be either 0 or 1, then afterward the two bits are either both 0 or both 1. The controlled-NOT operation causes the second bit, with zero bits of entropy initially, to become correlated with the state of the first bit, so that the second bit's final entropy is one bit. The ignorance in the first bit infects the second, causing its entropy to increase.

Figure 7. Pool and Loschmidt's Paradox

7a. *Loschmidt pointed out that if one reverses the velocities of the molecules in a gas, or of the balls in a pool table, entropy can apparently decrease. In figure 7a, the balls are in the same positions as in figure 6b, but their velocities been reversed.*

7b. *After their velocities are reversed, in the absence of friction the subsequent collisions of the balls undo their previous collisions, returning the balls to their original, low-entropy arrangement.*

To undo the controlled-NOT, simply perform it a second time. After the first operation, the two bits are either both 0 or both 1. During the second controlled-NOT op, if the control bit is 0 the second bit remains 0. If the control bit is 1, then the second bit is flipped from 1 to 0. In either case, the second controlled-NOT undoes the effect of the first and restores the second bit to 0. As a result, the bit's entropy decreases from one to zero bits.

Another implementation of Loschmidt's proposal is the spin-echo effect. To understand the spin-echo effect, consider the following macroscopic analogy. Runners line up at the starting line of a racetrack. When the gun goes off, the runners start running laps around the track. Because they run at different speeds and some are on the inside and some on the outside, after a few laps they are spread out all around the track. After ten minutes, a second gun is fired. At the sound of the second gun, the runners turn around and start running in the opposite direction. If each runner runs at the same speed as before, they gradually begin to bunch together as they undo the earlier distance. After ten minutes they all reach the starting line together.

In the spin-echo effect, the runners are nuclear spins. The protons and neutrons that make up the nuclei of atoms spin like little tops, with spin conventionally defined as either "up" or "down," the difference being determined by the direction of the spin: if you imagine a clock sitting faceup on a table, "spin up" is counterclockwise and "spin down" is clockwise. A convenient way to think of "up" or "down" spin is to curl the fingers of your right hand in the direction that the proton or neutron is spinning. Your thumb will then point along the axis about which it spins, and the direction your thumb is pointing in defines the "direction" of spin—that is, up or down.

Consider a bunch of protons all initially spinning in the same direction. If their spins are known, the entropy of each is zero. Now a microwave pulse sets the spins all precessing at once. (Precession is the wobbling motion a top makes under the force of gravity; nuclear spins are like little tops that wobble under the force of magnetism.) Each spin precesses at a slightly different rate and soon the spins are pointing in all directions, like the runners spread out in all positions on the track. The rate at which each spin precesses is given by its local magnetic field; this rate is "invisible" information, inaccessible to a macroscopic observer. Since the directions the spins are pointing in are now unknown, the spins, on their own, now have a high, almost maximum entropy, equal to the number of bits required to specify their direction of spin (that is, up or down) to the accuracy allowed by quantum mechanics.

The increase in entropy of the individual spins is an example of entropy increase by the spread of information. The precessing spins have become "infected" by the information in the local magnetic field. If we possessed that information, we could infer the directions in which the spins are pointing. As it is, we don't have that information, and since the spins are becoming correlated with the magnetic field, their individual entropies are increasing.

Now comes the echo. A second microwave pulse inverts the angles that the spins have precessed; for example, an angle of +60° becomes an angle of −60°. Now as each spin precesses, it undoes the angle precessed previously. After the same amount of time it took the spins to become unknown, the spins are once again all pointing in the same direction. Their entropy has decreased back to zero.

The spin-echo effect was first demonstrated experimentally fifty years ago. There are more complicated analogs of Loschmidt's notion, but all such analogs boil down to the same procedure. If you are a sufficiently skilled experimentalist and Boltzmann says, "Reverse them," you can reverse them!

Why does the spin-echo effect not constitute a violation of the second law of thermodynamics? The second law says that increases in entropy cannot be reversed. In the case of the spin-echo effect, entropy has only *apparently* increased. Even though the entropy of the spins taken on their own increases and then decreases during the course of the echo, the underlying entropy of the spins *taken together with the magnetic field* remains the same.

Exorcising Maxwell's Demon

There is a second way in which entropy can decrease. Recall that entropy is information that is unknown (that is, invisible). What happens when previously unknown information becomes known, when the invisible becomes visible? What happens when you get information? Then entropy decreases.

This mode of entropy decrease was first identified by James Clerk Maxwell, whose demon operates by getting information about the microscopic state of a volume of gas and is therefore able to decrease its entropy. Maxwell's demon has provoked many efforts to exorcise it over the years. The full exorcism of the demon was not accomplished until recently. (I played some part in this ceremony myself.) Despite the confusion sown by Maxwell's demon over the years, the final resolution is surprisingly simple: The underlying laws of physics preserve information. As a result, the total information/entropy of the gas and demon *taken together* cannot decrease.

In practice, this simple resolution involves considerable subtlety. Later, I'll present a full quantum-mechanical model of Maxwell's demon that will explain in detail the way in which the demon gets information and does its work. For the moment, consider a simple bit model, like those already discussed. Take two bits. The first bit, corresponding to the demon, reads 0 initially. The second bit, corresponding to the gas, can read either 0 or 1. The demon has no bits of entropy initially, and the gas has one bit.

The first step in the process of extracting entropy is for the demon's bit to get information about the gas bit. This can be accomplished by performing a controlled-NOT operation on the demon's bit with the gas bit as the control. The controlled-NOT flips the demon's bit if and only if the gas bit is 1. As a result, after the operation the demon's bit reads the same as the gas bit: they are either both 0 or both 1. That is, the demon's bit and the gas bit possess one bit of mutual information. The demon's bit has, in effect, measured the state of the gas bit to obtain that mutual information.

The second step of the process asks the demon to reduce the entropy of the gas. The demon can do this by performing a controlled-NOT operation on the gas bit using his own bit as control. Since the two bits are the same, the second controlled-NOT restores the gas bit to 0. If the demon's bit is 0, he leaves the gas bit in the state 0. If the demon's bit is 1, he flips the gas bit from 1 to 0. In either case, the gas bit is now in the state 0 and has zero bits of entropy. The demon has reduced the entropy of the gas by one bit.

The final situation is as follows. The gas bit is in the state 0. The demon's bit is 0 if the gas bit was initially in the state 0, and is 1 if the gas bit was initially in the state 1. The two controlled-NOT operations have in effect swapped the initial bit of the demon with the initial bit of the gas. Even though the entropy of the gas has decreased by a bit, the total amount of information in the gas and the demon taken together remains constant. The demon does not violate the second law of thermodynamics.

Note that the transfer of information from the gas to the demon takes place in compliance with Landauer's principle, discussed earlier. The demon's goal is to "erase" the gas bit by restoring it to 0. But because the underlying laws of physics preserve information, he can restore the gas bit to 0 only by transferring the information in the gas bit to his own bit. The total information remains constant.

In a *Scientific American* article on Maxwell's demon, Charles Bennett of IBM showed how Landauer's principle prevents the demon from violating the second law of thermodynamics in extracting work from a one-particle gas.* In a paper published in *Physical Review,* I showed that this

---

*Charles H. Bennett, "Demons, Engines, and the Second Law," *Scientific American* 257, no. 5 (November 1987): 108–16.

argument applies not only to systems of bits but to all physical systems—heat engines, hurricanes, what have you.* Physical dynamics can be used to get information, and that information can be used to decrease the entropy of a particular element of a system, but the total amount of information/entropy does not decrease. (The reader who is interested in demonology is directed to Harvey Leff and Andrew Rex's two compendia of papers on Maxwell's demon.)

If the resolution of the Maxwell's demon problem rests simply on the foundations of the information-preserving character of physical law, why has this problem caused so much confusion over the last century and a half? The problem lies with the distinction between information and entropy. Recall that entropy is invisible information, or ignorance—information that is unavailable. But the distinction between "visible" and "invisible" depends on who is looking. It is in fact possible to decrease entropy by looking at something.

To see how the distinction between visible and invisible information plays out in the case of Maxwell's demon, compare the perspective of a demon with that of an outsider. Like the demon, the outsider knows that the demon's bit is originally 0 but does not know the initial value of the gas bit. Unlike the demon, the outsider cannot follow the results of the sequence of controlled-NOT operations. She knows only that this sequence of controlled-NOT operations is taking place. That is, the outsider and the demon agree on the dynamics of the interaction between the demon's bit and the gas bit but draw the line between visible and invisible in different places. The demon's bit after the first op is visible to the demon but not to the outsider.

Before the first controlled-NOT, the demon and the outsider agree that the entropy of the demon's bit is zero and the entropy of the gas bit is one. After the first controlled-NOT, the demon's bit is perfectly correlated with the state of the gas bit. That is, the demon now "knows" the

---

*"Use of Mutual Information to Decrease Entropy: Implications for the Second Law of Thermodynamics," *Physical Review A* 39 (1989): 5378–86.

value of the gas bit. More precisely, as far as the demon is concerned, the entropy of the gas bit is zero, because it is conditioned on the state of the demon's bit after the operation, a state the demon knows. The information in the gas bit has gone from being invisible to the demon to being visible. As far as the demon is concerned, entropy has decreased by one bit, and visible information has increased by one bit.

Now consider the perspective of the outsider. After the first controlled-NOT, the outsider knows that the demon's bit and the gas bit are perfectly correlated. They read either 00 or 11, but the outsider does not know which. Accordingly, the outsider considers the demon's bit and the gas bit taken together to have one bit of entropy. Since the information in the gas and the demon remains invisible to the outsider, the outsider considers the entropy to have remained constant, with a value of one bit.

After the second controlled-NOT, the bit that was originally in the gas has been transferred to the demon. Both demon and outsider agree that the gas has been restored to the state 0. From the demon's perspective, his bit is visible and registers one bit of information: the entropy is zero. From the outsider's perspective, the demon's bit is invisible: the entropy is one bit. Both demon and outsider agree that the total amount of information is one bit. The second law of thermodynamics applies to the total amount of information, both visible and invisible.

Maxwell's demon completes the discussion of entropy increase and entropy decrease for the moment. At bottom, as the statistical mechanicians of the late nineteenth century showed, the world is made up of bits. The second law of thermodynamics is a statement about information processing: the underlying physical dynamics of the universe preserve bits and prevent their number from decreasing. To fully understand these physical dynamics requires us to look at quantum mechanics, which describes how physical systems behave at their most fundamental level. Before turning to quantum mechanics, however, let's look briefly at the information-processing capacity of classical systems, such as atoms in a gas or snooker balls on a table.

Atomic Computation

The position and velocity of an atom in a gas register information. Indeed, positions and velocities of atoms were the very first quantities to which the basic formulas of information were applied. Atoms register bits.

What about processing that information? When two atoms in a gas collide, the information they register is transformed and processed. How does the information processing performed during atomic collision relate to the information processing performed by the logic gates described in the first part of this book? In fact, as pointed out by Edward Fredkin of Carnegie Mellon University and Tommaso Toffoli of Boston University, atomic collisions naturally perform AND, OR, NOT, and COPY logic operations. In the language of information processing, atomic collisions are computationally universal.

In Fredkin and Toffoli's model, each possible atomic collision performs AND, OR, NOT, or COPY operations on suitably defined input and output bits. By assigning the proper initial positions and velocities to atoms in a gas, it is a straightforward matter to "wire up" any desired logic circuit. Atoms bouncing in a gas are, in principle, capable of universal digital computation.

In practice, of course, it is rather difficult to make a gas of atoms perform a computation. Even if we did have control over the position and velocity of individual atoms, quantum mechanics limits the accuracy to which position and velocity can be simultaneously specified. Moreover, the collisions between atoms in a gas is intrinsically chaotic; that is, a slight error in the specification of the initial positions and velocities of atoms will typically grow in time, via the butterfly effect, until it contaminates the entire computation. As will be seen in the following chapters, though, both of these limitations can be overcome by using more suitable quantum-mechanical systems to perform computation.

Although practical limitations prevent using collisions between atoms in a gas to compute, the fact that atomic collisions in principle allow computation implies that the long-term future of a gas of atoms is intrinsically unpredictable. The halting problem (see chapter 2) foils not only conventional digital computers but any system capable of performing digital logic. Since colliding atoms intrinsically perform digital logic, their long-term future behavior is uncomputable.

This computational capacity of colliding spheres throws light on the possibility of a third demon, this one evoked by the Marquis Pierre-Simon de Laplace. In an essay on using Newtonian mechanics to predict the future behavior of heavenly bodies, Laplace wrote:

> We may regard the present state of the universe as the effect of its past and the cause of its future. An intellect which at any given moment knew all of the forces that animate nature and the mutual positions of the beings that compose it, if this intellect were vast enough to submit the data to analysis, could condense into a single formula the movement of the greatest bodies of the universe and that of the lightest atom; for such an intellect nothing could be uncertain and the future just like the past would be present before its eyes.

A being capable of performing this prodigious task of prediction is called Laplace's demon.

Even if the underlying laws of physics were fully deterministic, however, the computational ability of simple systems such as colliding spheres implies that to perform the type of simulation Laplace envisaged, the calculating demon would have to have at least as much computational power as the universe as a whole. Since, as we shall see, computational power requires physical resources, Laplace's demon would have to use at least as much space, time, and energy as the universe itself.

A second problem with Laplace's demon is that the laws of quantum mechanics are not deterministic in the sense required by Laplace. In quantum mechanics, what happens in the future is predictable only in a probabilistic way. In fact, the motions of heavenly bodies are intrinsically chaotic and thus are constantly pumping information up from microscopic to macroscopic scales. As will be shown in the next chapter, the result of this celestial chaos is that even Laplace's heavenly bodies move in a probabilistic fashion that cannot be predicted, even by a demon.

CHAPTER 5

Quantum Mechanics

In the Garden

I was standing in the garden of the Master's Lodge at Emmanuel College, Cambridge, sipping a glass of champagne. It was the spring of 1983. My fellow graduate students and I were talking about the usual stuff of Cambridge life: boat races, May balls, and the upcoming mathematical tripos examinations that would determine our future. A stunning older woman interrupted us. "You fools!" she exclaimed, in a pronounced Spanish accent. "Don't you see that the world's greatest author is sitting over there with no one to talk to?" I looked where she was pointing and saw an old blind man in a white suit sitting quietly on a bench. It was Jorge Luis Borges, and the woman was his companion, Maria Kodama. She shepherded us over to him.

There was a question I had always wanted to ask Borges, and at last I had the opportunity. In his story "The Garden of Forking Paths," Borges envisions a world in which all possibilities actually happen. At each decision point, each fork in the path, the world takes not one alternative but both at once. So, Borges writes:

> In the work of Ts'ui Pên, all possible outcomes occur; each one is the point of departure for other forkings. Sometimes, the paths of this labyrinth converge: for example, you arrive at this house, but in one of the possible pasts you are my enemy, in another, my friend. . . . [Ts'ui Pên] did not believe in a uniform, absolute time.

He believed in an infinite series of times, in a growing, dizzying net of divergent, convergent and parallel times. This network of times which approached one another, forked, broke off, or were unaware of one another for centuries, embraces *all* possibilities of time. We do not exist in the majority of these times; in some you exist, and not I; in others I, and not you; in others, both of us.

"Dr. Borges," I said, "when you wrote your story, were you aware that it mirrors the so-called Many Worlds interpretation of quantum mechanics? In this interpretation, whenever anyone makes a measurement that reveals information about the world being one way or another, the world splits in two and takes both paths. In the conventional interpretation of quantum mechanics, the Copenhagen interpretation, if I ask a nuclear particle whether it is spinning clockwise or counterclockwise, it picks one spin or the other with equal probability. But in the Many Worlds interpretation, at the moment of measurement the world's path forks and it takes not one fork or the other but both at once."

Borges asked me to repeat the question in a more comprehensible fashion. When he understood that I was asking whether or not the foundations of quantum mechanics had influenced his writing, he answered, "No." He went on to say that although he had not been influenced by work on quantum mechanics, he was not surprised that the laws of physics mirrored ideas from literature. After all, physicists were readers, too.

And in fact "The Garden of Forking Paths" was published in 1941, years before John Wheeler's student Hugh Everett introduced the Many Worlds interpretation of quantum mechanics. So if there was influence, it was from literature to physics, not the other way around.

Wave-Particle Duality

Quantum mechanics is in fact very much like one of Borges's *Ficciones*. But its *weirdness*, as I called it earlier, accurately reflects the fundamental structure of the universe. In the early days of the twentieth century, the

Danish physicist Niels Bohr used quantum mechanics to successfully explain the structure of the hydrogen atom, but—like most workers in quantum mechanics, including Einstein—Bohr found the theory counterintuitive. Einstein's response was to reject quantum mechanics ("God does not play dice," he said). Bohr's, on the other hand, was to develop an almost mystical philosophy of the quantum world. Whatever attitude you choose to adopt toward quantum mechanics, though, if you get dizzy contemplating it, that's a good sign. Of course, dizziness in itself doesn't guarantee that you have understood quantum mechanics, but it's a start.

To attain an understanding of quantum mechanics at the intuitive (or, more precisely, the counterintuitive) level, a good start is to contemplate the principle that Bohr called wave-particle duality. "Wave-particle duality" refers to the fact that things we normally think of as waves, like light or sound, are actually made up of particles, or quanta (*quantum* is the Latin word for "how much"). Particles of light are called photons ("-on" being the usual suffix denoting a particle); particles of sound are called phonons.

A simple experiment demonstrates the quantum nature of light. A photodetector is a device that detects light. It produces an electrical current whose magnitude is proportional to the amount of light it absorbs. A photodetector in a bright room produces a lot of current. When the light is turned down, the current goes down, too. When the lights are turned off and the shades are drawn, the current gets close to zero. Finally, if one takes pains to exclude almost all light, covering windows and doors with plywood and taping all the cracks, the photodetector starts to exhibit a different sort of behavior. The current is zero most of the time, but every now and then it spikes in a short burst. The photodetector is detecting *individual photons.*

The notion that waves are actually made of particles is very old. That sound is made of waves has been known since Pythagoras, but the ancient Greeks thought light was composed of particles and argued over whether these particles came from the eye or from the object viewed. Newton proposed a "corpuscular theory" of light in terms of particles. But Newton's own experiments with prisms were more easily explained if light was made of waves, and the wave theory of light dominated from the seventeenth century until the end of the nineteenth, culminating in Maxwell's equations, which explained all known electromagnetic phenomena in terms of waves of light.

There was a problem with thinking of light only in terms of waves, however, and that problem had to do with heat. At the end of the nineteenth century, Max Planck analyzed the light emitted from a stove that was heated until it glowed red. Such light is called "blackbody radiation." Black objects absorb and emit light of all frequencies (frequency being the rate at which a light wave wiggles up and down). Planck pointed out that if light was indeed made of waves, then the amount of energy and entropy in the radiation emerging from hot objects should be infinite, a serious problem for both the first and second laws of thermodynamics. He was able to resolve this problem by assuming that light was made out of particles whose energy was proportional to the frequency of the wave. Planck called these particles photons. Each photon carried a small amount of energy—a quantum. Planck found that if the energy of each of these particles (measured in joules) was equal to 6.63×10^{-34} times the wave's frequency per second, then energy was conserved by the radiant heat. Planck's constant relates energy to frequency. It is so ubiquitous in physics that it has been given its own special symbol, h.

Things we think of as waves correspond to particles; this is the first aspect of wave-particle duality. The second, complementary aspect of wave-particle duality is that things we think of as particles correspond to waves. Just as every wave is made up of particles, every particle—an electron, an atom, a pebble—has a wave associated with it. The wave is associated with the position of the particle: the particle is more likely to be found in places where the wave is big. The distance between the peaks of the wave is related to the particle's speed: the smaller the distance from peak to peak, the faster the particle is going. Finally, the wave's frequency is proportional to the energy of the particle. In fact, the particle's energy is exactly equal to the frequency times Planck's constant.

The Double-Slit Experiment

The double-slit experiment demonstrates the wave nature of particles. Waves superpose, or *interfere,* with each other. If my daughter Emma, sitting at one end of the bathtub, sets a wave going toward her sister Zoe, sitting at the other end, and Zoe sets a wave going toward Emma at the same time, then when the waves meet in the middle, they splash out all over me. Light waves will combine to interfere with each other in the same way. If you shine a beam of light at a screen that has two slits in it and look at the pattern the light makes on the wall beyond the screen, you see alternating bands of light and dark. This is called an "interference pattern." The light waves, moving like water waves through pilings, go through both slits at once; each wave thus splits in two and then combines on the wall. The bands of light occur in places where the peaks and troughs of the wave from one slit coincide with the peaks and troughs of the wave from the other and reinforce them, a phenomenon called "positive interference." The bands of dark occur where the peaks of the wave from one slit coincide with the troughs of the other wave and the two waves cancel each other out, a phenomenon called "negative interference." If you cover one of the slits, the interference pattern goes away, because there is no wave to interfere with the wave that goes through the remaining slit. Interference, crucially, requires the wave to go through both slits at once.

The double-slit experiment can be performed with particles as well. Shoot a beam of particles—electrons, say—at a screen with two slits and put a photographic plate on the wall to record where the particles land. Each particle leaves a spot on the photographic plate. If you close up the

Figure 8. The Double-Slit Experiment

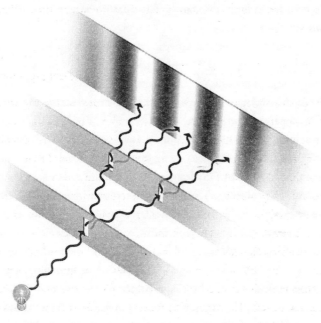

In the double-slit experiment, particles are sent first through a single slit, then through a double slit, before landing on a screen. The pattern that the particles make on a screen exhibits an "interference pattern," evidence for the underlying wavelike nature of the particles.

left slit and let the electrons go only through the right slit, then you get one pattern of spots. If you close up the right slit and let the electrons go only through the left slit, you get another pattern.

Now open both slits. What sort of pattern do you expect to see? Classical reasoning says that each electron will go through either one slit or the other. Thus you expect to see a distribution of spots on the photographic plate that is just the two single-slit distributions of the previous paragraph combined. You don't expect to see an interference pattern, because each of the particles should go through only one slit. An interference pattern is a wave phenomenon; it arises because waves can and do go through both slits at once. But a particle is a particle: it can't go through both slits at once.

Perform the double-slit experiment with particles. What do you see? An interference pattern! The spots made by the individual particles fall across the photographic plate in a series of bands. When you cover one of the slits, the interference pattern disappears. Evidently, the particles behave as if they were waves.

What's going on? Maybe the electrons going through one slit are knocking into the electrons going through the other slit and making some kind of pattern. OK, decrease the number of electrons in the beam to minimize collisions. The interference pattern remains. Now shoot a single electron at the screen. The interference pattern is still there, but now it simply governs the probability for where the electron arrives on the photographic plate: it preferentially lands where the band of spots lies, thus the interference pattern can't be caused by multiple particles interfering with one another. There is only one electron, but somehow it still behaves as if it were a wave. The experiment reveals that the particle goes through both slits at once. An electron, a proton, a photon, an atom can be in two places at the same time.

The double-slit experiment illustrates the fact that a particle doesn't have to be either "here" or "there." Because of its underlying wavelike nature, a particle can be both "here" and "there" at the same time. This ability of things to be in many places at once is responsible for the power of quantum computation, which we will explore further later on.

Decoherence

If things can be in two places at once, then why don't we see pebbles, people, and planets showing up in more places than one? The Austrian physicist Anton Zeilinger has performed the double-slit experiment successfully with so-called buckyballs, soccer-ball-like constructions made of sixty carbon atoms. He plans next to do the double-slit experiment with bacteria about 100 times larger. The bigger something is, however, the harder it is to coax it into existing in two places at once. (Big things tend to behave more "classically," and less quantum-mechanically.) The reason lies not so much with the physical size of the object as with its visibility. The bigger something is, the more interactions it tends to have with its surroundings, thus the easier it is to detect. *In order to go through both slits at once and produce an interference pattern, a particle must pass through the slits undetected.*

Suppose you place a detector on the right-hand slit. The detector registers the presence or absence of a particle at the slit, letting the particle pass through otherwise unchanged. When the detector detects a particle, it clicks. Now perform the double-slit experiment with the detector operating. Look at the screen. The interference pattern has disappeared!

What happened? Recall that an interference pattern stems from the wave associated with a particle. That wave naturally goes through both slits at once. When the detector is operating, a particle going through the right slit will cause it to click. A particle going through the left slit will not. (Whether or not the detector clicks is random: the particle goes through one slit or the other with equal probability.)

When the detector clicks and detects the particle, the wave corresponding to the particle has been localized to the right slit. When the detector fails to click, the particle has been localized to the left slit. This process of localization of the wave is sometimes called "collapse of the wave function." That is, when the detector is "observing" the right slit, the particle has to go through either one slit or the other; it no longer goes through both slits at once. And since the wave corresponding to the particle no longer goes through both slits at once, it cannot interfere with itself to produce the interference pattern's alternating bands of light and dark.

Observation (or measurement, as it is conventionally called) destroys interference. Without measurement, the particle merrily goes through both slits at once; with measurement, it goes through one or the other. In other words, measurement intrinsically disturbs the particle. When you ask the particle where it is, it is forced to confess that it is in one place or another and no longer in both places at once.

It is interesting to note, in the above experiment, that the measurement disturbs the particle's wave whether or not the detector clicks. The detector clicks only if the particle goes through the right-hand slit, where the detector is located. But when the detector fails to click, meaning that the particle has gone through the left-hand slit, the interference pattern is still destroyed—that is, the measurement still disturbs the particle's wave. The particle need not ever come close to the detector. (Are you dizzy yet?) Nor does the detector have to be a macroscopic device: All that is required to destroy the interference pattern is for some system, no matter how small, to get information about the position of the particle. If the particle knocks a passing electron or molecule of air, for example, that, too, will destroy the interference pattern.

It is now clear why big things tend to show up in one place or another, but not both. Pebbles, people, and planets are constantly interacting with their surroundings. Each interaction with an electron, a molecule of air, a particle of light tends to localize a system. Big things interact with lots of little things, each of which gets information about the location of the big thing. As a result, big things tend to appear here or there instead of here and there at the same time.

The process by which the environment destroys the wavelike nature of things by getting information about a quantum system is called "decoherence." Decoherence is a common process. Remember the argument given earlier for entropy increase: almost any interaction between one thing and another causes the first thing to get information about the second, and vice versa. As the spread of ignorance shows, these interactions cause the entropies of the things taken on their own to increase. The same mechanism operates to make quantum objects behave in a more classical way.

Quantum Bits

In the previous chapter, each mechanism by which information was conserved, spread about, erased, or increased was illustrated by a simple example presented in terms of bits. To understand how quantum mechanics work, then, wouldn't it be nice to have a similar, quantum-mechanical device? A good example of a quantum-mechanical bit, or qubit, is a nuclear spin, like that of protons and neutrons in the spin-echo effect. "Spin up" is conventionally given a bit value of 0 and "spin down" a value of 1. The bit value of a nuclear spin can be determined by putting the spin through a device called a Stern-Gerlach apparatus, which discriminates between 0 and 1 by moving spin-up nuclei in one direction and spin-down nuclei in the opposite direction (their positions are recorded on a photographic plate). Both possible values for spin correspond to waves: a wave moving counterclockwise for spin-up (or 0) and a wave moving clockwise for spin-down (or 1). The wave corresponding to 0 is customarily represented by the symbol $|0\rangle$ and the wave corresponding to 1 by the symbol $|1\rangle$. The "$|\ \rangle$" or "bracket" notation has a mathematical significance, but for our purposes here it simply serves to indicate that whatever appears within the brackets is a quantum-mechanical object—a wave.

It is possible to combine waves. The resulting combination is referred to as a "superposition." What is the state of the system corresponding to the sum—or superposition—of the wave for spin-up and the wave for spin-down? That is, what wave corresponds to the state $|0\rangle + |1\rangle$? In the case of spins, this state turns out to be easy to visualize: it is a state of spin along an axis perpendicular to the axis defining spin-up and spin-down. Spin-up plus spin-down is spin sideways!

It's also possible to subtract waves from each other. The wave designated $-|1\rangle$ is a wave whose troughs correspond to the peaks of the wave $|1\rangle$ and whose peaks correspond to the troughs of the wave $|1\rangle$. That is, $-|1\rangle$ wiggles where $|1\rangle$ waggles, and vice versa. Now look at the superposition $|0\rangle - |1\rangle$. This state is also easily visualized. It is a state of spin

Figure 9abc. Quantum Bits

A nuclear spin is a quantum bit. Spin counterclockwise, or "up," registers the logical state 0 (figure 9a). Spin clockwise, or "down," registers the logical state 1 (figure 9b). Spin "sideways" is a quantum state that registers 0 and 1 at the same time.

along the same axis as the state $|0\rangle + |1\rangle$, but in the opposite direction. Thus the direction of the spin depends crucially on the sign (or phase) of each wave in the superposition. We could distinguish between these states along the sideways axis by taking the Stern-Gerlach apparatus and turning it on its side.

The state $|0\rangle + |1\rangle$ has a definite value of spin along the sideways axis. If you measure which direction it is spinning about that axis, you always find that it is spinning clockwise. But when you take this same spin and try to determine its value of spin about the vertical axis, the result will be completely random; half the time you will find that it is counterclockwise (that is, you find the state spin-up, or $|0\rangle$) and half the time you will find that it is clockwise (spin-down, or $|1\rangle$). When the value of spin about the sideways axis is completely certain, the value of spin about the vertical axis is completely uncertain.

Similarly, the state $|0\rangle$ has a definite value of spin along the vertical axis. If you measure the spin, you find that it is clockwise (spin-up). But now the value of spin about the sideways axis is completely uncertain; if you measure the spin about the sideways axis, half the time you'll find it spinning clockwise about this axis and half the time you'll find it spinning counterclockwise. Now that the value of spin about the vertical axis is certain, the value of spin about the sideways axis is uncertain.

The Heisenberg Uncertainty Principle

Apparently it is not possible to have a definite value of spin about two different axes at the same time. This intrinsically chancy nature of quantum mechanics was immortalized by Werner Heisenberg, one of the founders of quantum mechanics, as the "uncertainty principle." The uncertainty principle states that *if the value of some physical quantity is certain, then the value of a complementary quantity is uncertain.* Spin about the vertical axis and spin about the sideways axis are just such complementary quantities: if you know one, you can't know the other.

Another pair of complementary quantities are position and speed: if you know the position of a particle exactly, then you know nothing of how fast it's going. (Traffic cop pulls over Heisenberg's car: "Professor Heisenberg, do you have any idea how fast you were going?" Heisenberg: "No, but I know exactly where I am.")

The Heisenberg uncertainty principle expresses a trade-off between the degree of certainty about the value of one physical quantity, such as position, and a complementary quantity, such as speed. The more certain the value of one quantity, the less certain the value of the other. As a consequence, any procedure (such as measurement, or observation) that makes the value of some physical quantity more precise inherently makes the value of the complementary quantity less precise. *Again, measurement tends to disturb the system measured.*

This perturbing aspect of the Heisenberg uncertainty principle has become deeply embedded in popular lore. For example, the uncertainty principle is sometimes invoked, incorrectly, to explain why anthropologists intrinsically alter the societies they set out to investigate. (As in the expression, "When the anthropologist comes in the door, the truth flies out the window.") In fact, the Heisenberg uncertainty principle typically makes a difference only at very small scales, such as the atomic scale. Even the most probing investigations by anthropologists occur in a realm much too large to allow the uncertainty principle to kick in.

Flipping Qubits

It's not hard to flip a quantum bit, or qubit. Recall from the example of the spin-echo effect that when you put a nuclear spin in a magnetic field, the spin precesses about that field. Take a spin that's initially spin-up (or |0⟩) and apply a field facing toward you. After half the time it takes for the spin to precess all the way around, it has precessed to the spin-down state, |1⟩. (Likewise, if the spin starts out down, or |1⟩, after the same amount of time it will precess to the spin-up state, |0⟩.) By applying the magnetic field, you flip the qubit.

By varying the amount of time for which you apply the magnetic field, you can also put the spin in a variety of superpositions. For example, start with the spin in the state spin-up and apply the field for one-quarter of the time it takes to precess all the way around; the spin is now in the state spin sideways to the right, or $|0\rangle + |1\rangle$. Or start with the spin-up state and apply the field for three-quarters of the time it takes to precess all the way around; the spin is now in the state spin sideways to the left, or $|0\rangle - |1\rangle$. By applying the magnetic field for different amounts of time, you can rotate the spin into any desired superposition of states.

These single-qubit rotations are the quantum analogs of single classical bit transformations, such as bit-flip, or NOT. Because of the existence of superpositions, there are many more transformations that can be applied to a quantum bit than to a classical bit. One thing classical and qubit transformations have in common, though, is that each is a one-to-one transformation. It's straightforward to reverse the action: just rotate the qubit back along the same axis in the opposite direction. Like the transformations allowed by classical physics, the rotations of a qubit preserve information.

Now look at interactions between qubits. Consider a two-qubit transformation that is a quantum analog of the controlled-NOT logic operation described earlier. Recall that the controlled-NOT operation flips one bit if and only if the other bit, the control bit, is 1. That is, the controlled-NOT operation takes 00 to 00, 01 to 01, 10 to 11, and 11 to 10. Here the first bit is the control bit. The controlled-NOT operation is one-to-one and can be reversed simply by applying it twice. The quantum controlled-NOT takes the quantum states $|00\rangle$ to $|00\rangle$, $|01\rangle$ to $|01\rangle$, $|10\rangle$ to $|11\rangle$, and $|11\rangle$ to $|10\rangle$. Here, the state $|00\rangle$ corresponds to the "joint wave" of the two quantum bits taken together, in which the first qubit is in the state $|0\rangle$ and the second qubit is also in the state $|0\rangle$.

The preceding paragraphs form the basis for quantum computation. Later, we'll see that rotations of individual quantum bits, together with controlled-NOT operations, constitute a universal set of quantum logic operations. Recall that AND, OR, NOT, and COPY make up a universal set of classical logic operations; any desired logical transformation can be built up from these basic elements. Similarly, any desired transformation of a set of quantum bits can be built up out of single-qubit rotations and controlled-NOTs. This universal feature can be used to perform arbitrarily complicated quantum computations. But first, let's use the universal character of rotations and controlled-NOTs to investigate how processes such as measurement and decoherence actually work.

Qubits and Decoherence

The state $|0\rangle + |1\rangle$ is a qubit analog of the state of the particle in the double-slit experiment in which it is going through both slits at once. The state of the particle going through the slits also corresponds to a quantum bit. If $|left\rangle$ corresponds to the state in which the particle goes through the left-hand slit and $|right\rangle$ corresponds to the state in which the particle goes through the right-hand slit, then $|left\rangle + |right\rangle$ is the state in which the particle goes through both slits at once.

A qubit such as a nuclear spin can be placed in the state $|0\rangle + |1\rangle$ (corresponding to the particle going through both slits at once) by preparing a spin in the state spin-up ($|0\rangle$) and rotating it one-quarter of a turn, into the state $|0\rangle + |1\rangle$. Similarly, you can verify that the qubit is in the desired state by rotating the spin back one-quarter of a turn and measuring its state (e.g., with a Stern-Gerlach apparatus) to verify that it has been returned to its original state.

Now take a second qubit, initially in the state $|0\rangle$. Just as the first qubit is an analog to the position of the particle, this second qubit is an analog to the detector. Perform a controlled-NOT op on this qubit, using the particle bit as the control. The controlled-NOT flips the detector qubit if and only if the particle qubit is in the state $|1\rangle$, corresponding to the particle going through the right-hand slit. But this particle qubit is in the superposition state $|0\rangle + |1\rangle$. The *quantum* controlled-NOT operation acts like a classical controlled-NOT *on each component of this superposition*. In the part of the superposition in which the particle qubit is in the state $|0\rangle$, corresponding to the particle going through the left-hand slit, the detector qubit remains in the state $|0\rangle$. In the part of the superposition in which the particle qubit is in the state $|1\rangle$, the detector qubit is flipped from $|0\rangle$ to $|1\rangle$. Taken together, after the controlled-NOT, the two quantum bits are now in the state $|00\rangle + |11\rangle$. In one component of the superposition, particle and detector qubits are both $|0\rangle$. In the other component, they are both $|1\rangle$. The controlled-NOT operation *correlates* the two quantum bits.

In the course of the controlled-NOT operation, information in the first qubit has spread to and "infected" the second qubit; that is, the controlled-NOT operation has created mutual information between the two qubits. The second qubit now possesses information about whether the first qubit is $|0\rangle$ or $|1\rangle$.

The controlled-NOT operation has also disturbed the first qubit. Suppose you try to verify that the first qubit is still in the state $|0\rangle + |1\rangle$ by rotating the nuclear spin back a quarter of a turn and measuring to see if it is in the state spin-up. When you make this measurement, you find that half the time the spin is in the correct state, spin-up, and half the time it is in the incorrect state, spin-down. The particle qubit is no longer in the state $|0\rangle + |1\rangle$. In the process of correlating the particle qubit with the detector qubit, the controlled-NOT operation has completely randomized the state of the particle qubit.

Like its classical counterpart, the quantum controlled-NOT allows one bit to obtain information about another. But unlike its classical counterpart, the quantum controlled-NOT typically disturbs the bit about which information is obtained. This disturbance is intrinsic to processes in which one quantum system gets information about another; in particular, the quantum measurement process typically disturbs the system measured.

In the example given here, the disturbance can be undone simply by repeating the controlled-NOT operation. Like the classical controlled-NOT, the quantum controlled-NOT is its own inverse. If you perform it twice, you return the qubits to their original state. In particular, a quantum controlled-NOT performed on the state $|00\rangle + |11\rangle$, with the first qubit as control, does nothing to the $|00\rangle$ component of the superposition and takes the $|11\rangle$ component to $|10\rangle$. The second (detector) qubit is now in the state $|0\rangle$ and the first (particle) qubit is now in the state $|0\rangle +$ $|1\rangle$. Rotating the particle qubit back by a quarter of a turn and measuring gives the result spin-up, verifying that the particle qubit was indeed returned to the proper state.

Historically, though, the quantum measurement process is taken to be irreversible. Unlike the simple controlled-NOT model of quantum detection given here, conventional interpretations of quantum mechanics, such as Bohr's Copenhagen interpretation, assume that once a macroscopic measuring apparatus has become correlated with a microscopic system such as a particle, that correlation cannot be undone. In this purported irreversibility of measurement, the reader probably detects an echo of the second law of thermodynamics. In Boltzmann's H-theorem, you'll recall, the apparent irreversibility of entropy increase holds only as long as atoms don't interact in such a way as to undo their correlations and thus decrease their entropies. Similarly, in the quantum measurement process, irreversibility can only be apparent.

In particular, the underlying dynamics of quantum systems preserve information, just as the dynamics of classical systems do. Because these dynamics preserve information, they can in principle be reversed. Thus there is a quantum-measurement analog of Loschmidt's objection. Simply reverse the dynamics of the measurement process and the quantum system will be returned to its pristine, undisturbed state. As with the classical controlled-NOT discussed earlier, the second application of the quantum controlled-NOT operation is a realization of Loschmidt's objection. Analogs of the spin-echo experiment can effectively reverse the dynamics of millions of qubits at once.

To the (correct) objection to the notion of irreversibility in the measurement process, Bohr might well have replied, like Boltzmann, "Go ahead, reverse it." Niels Bohr, however, was a gentle person. His response to these objections was instead to obscure the problem of irreversibility by veiling the traditional Copenhagen interpretation in a semantic fog, wisps of which persist to the present day.

In fact, the idea of the irreversibility of quantum measurement is just as safe as the second law of thermodynamics, true or otherwise. Recall that with the second law, you identify an increase in the entropy of a system by, in effect, betting that the newly made correlations will not be reversed, thus undoing this apparent increase in entropy. If in fact those correlations are reversed, decreasing the entropy of the parts, then you lose your bet: entropy did not, in fact, increase.

Similarly, in the quantum-measurement process, you provisionally identify the spread of information from the system to the measurement apparatus as irreversible. If it turns out later that the dynamics of the measurement process undo themselves to restore the original state, you simply rescind your identification of the spread of information as an irreversible process. Since most of the time entropy continues to increase and information continues to spread, you rarely have to recant. Occasionally, however, because the laws of physics are reversible, an apparent entropy increase undoes itself and information "unspreads." Given the underlying reversibility of the known laws of physics, and the existence of phenomena like the spin-echo effect in which entropy does in fact decrease, you may find it conceptually more satisfying to regard the second law of thermodynamics and irreversibility of quantum measurements as *probabilistic* laws: entropy tends to increase and information is highly likely to spread. But sometimes they don't.

Entanglement

Another difference between the classical and quantum versions of the controlled-NOT is that in the quantum case, information is created, apparently from nothing. Recall the analogous classical process: the particle bit could have been in either the state 0 or the state 1 to begin with; it had one bit of entropy. Here, the qubit is in a well-defined state: its entropy is zero. Of course, the state that the qubit is in, $|0\rangle + |1\rangle$, is one that has elements of both $|0\rangle$ and $|1\rangle$: like the corresponding state of the particle in the double-slit experiment, this state is a curious quantum state in which the quantum bit in some sense registers 0 and 1 at the same time.

When the two classical bits interacted via the controlled-NOT, the entropy in the particle bit infected the detector bit. The two bits were now correlated and the entropy of the detector bit had increased. When the two qubits interact via the quantum controlled-NOT, they also become correlated, and the entropy of the detector qubit increases. *But this entropy did not come from the particle qubit.* In the quantum case, before the controlled-NOT was applied, the particle qubit was in a well-defined state with zero entropy. Where did the information come from?

What's going on is that *quantum mechanics, unlike classical mechanics, can create information out of nothing.* Take our two qubits in their correlated state, $|00\rangle + |11\rangle$, with the wave of the first qubit correlated with the wave of the second qubit. This state is a definite quantum state: its entropy is zero. But now each of the qubits on its own is in a completely indefinite state: each could be either $|0\rangle$ or $|1\rangle$. That is, each quantum bit now has one full bit of entropy.

This weird type of quantum correlation is called "entanglement." If a classical system is in a definite state, with zero entropy, then all the pieces of the system are also in a definite state, with zero entropy. If we know the state of the whole, then we also know the state of the pieces. For example, if two bits are in the state 01, then the first bit is in the state 0 and the second bit is in the state 1. But when a quantum system is in a definite state, though, such as the correlated state of our quantum bits, the pieces of the system *need not be in a definite state.* In entangled states, we can know the state of a quantum system as a whole but not know the state of the individual pieces.

When the pieces of a quantum system become entangled, their entropies increase. Almost any interaction will entangle the pieces of a quantum system. The universe is a quantum system, and almost all of its pieces are entangled. Later, we'll see how entanglement allows quantum computers to do things that classical computers can't do. Here, we see that entanglement is responsible for the generation of information in the universe.

Spooky Action at a Distance

Entanglement is responsible for what Einstein called "spooky action at a distance." Consider the state for two quantum bits $|01\rangle - |10\rangle$. In this state, if you look at the first qubit and find that it is 0, then the second qubit is 1. Similarly, if you look at the first qubit and find that it is 1, then the second qubit is 0. That is, the two qubits are the opposite of each other. For example, say the two qubits are made up of nuclear spins. When you measure the first spin along some axis and find that it is spin-up, the second spin will be spin-down.

So far, this doesn't sound so bad. The two spins rotate in the opposite direction, no matter which axis one chooses to measure their rotation about. The problem is that before the measurement of the first qubit, both qubits are in a completely indefinite state. Measuring the first qubit puts it in a definite state, $|0\rangle$ or $|1\rangle$. This is not surprising, as measurement is supposed to determine the state of the thing measured. What is surprising is that measuring the spin of the first particle about some axis also puts the *second* particle in a definite state of spin about that axis; that is, if you choose to measure the first spin about the vertical axis, then after the measurement the second spin is in a definite state of spin about the vertical axis. If you choose to measure the first spin about the sideways axis, then after the measurement the second spin is in a definite state about the sideways axis. Somehow, it appears that measuring the first spin does something to the second spin as well. And the first particle need be nowhere near the second particle. After entanglement, one particle could be kept here on Earth and the other particle sent to Alpha Centauri.

How can measuring something on Earth simultaneously affect something else at Alpha Centauri, which is some four light-years away from us? No signal can possibly arrive there in less than four years, let alone simultaneously. This is what Einstein meant when he called the effect of entanglement "spooky action at a distance." With Boris Podolsky and Nathan Rosen, he wrote a famous paper on what is now commonly referred to as the EPR paradox, pointing out the counterintuitive nature of entanglement and showing that it implied that there were no underlying "elements of reality" in the world.

In fact, entanglement does not involve action at a distance, spooky or otherwise. If measuring the spin of the first particle truly affected the spin of the second particle in some observable way, then it would be possible to send information from the first to the second by making measurements on the first. But measuring the first spin has no observable effect on the second spin. True, after the measurement of the first spin along the vertical axis, the second spin is in a definite state of spin about that axis: if the first spin was in the state $|0\rangle$, the second spin is in the state $|1\rangle$, and vice versa. But in the absence of information about the result of the measurement on the first spin, the second spin's state is still completely uncertain, just as it was before the first spin was measured. Measuring the first spin does not change the outcome of measurements made on the second spin; measuring the first spin has no observable effect on the second spin. Measuring the first spin may increase our knowledge of the second one, but it doesn't really change its state. As a result, it is not possible to send information from the first spin to the second just by making measurements on the first spin. Entanglement does not involve action at a distance.

Even if entanglement does not involve action at a distance, it is still spooky. Each spin registers one qubit, no more, no less. But the fact that two spins are always spinning in the opposite direction, no matter which axis one chooses to measure them about, seems to involve much more than one bit of information. As a classical analog, consider two brothers who, when given a choice between two alternatives, always pick the opposite choices. One brother walks into the Miracle of Science bar in

Cambridge, Massachusetts, at the same moment that the other brother walks into the Free Press pub in Cambridge, England. The bartender in the Miracle of Science asks the first brother, "Beer or whiskey?" "Beer," he replies. Meanwhile, the bartender in the Free Press asks the same question. "Whiskey," says the contradictory brother in Britain. If, instead, the bartenders had asked, "Bottled or draft?" one brother would have replied "Bottled" and the other "Draft." Or if the bartenders had asked, "Red wine or white?" one brother would have replied "Red" and the other "White." For each bit of information the bartenders might elicit, the brothers would reply with opposite bits.

There is nothing impossible about this kind of oppositeness. It is just that in the classical world the brothers must share a bit of information for each possible question that can be asked. In the quantum version of this story ("Two entangled spins walk into a bar . . ."), the two entangled spins share one and only one quantum bit, and yet they are capable of giving opposite answers to an infinite variety of questions corresponding to the infinite set of possible axes about which they can be measured. Spooky.

The Quantum-Measurement Problem

Quantum measurement is a process during which one quantum system gets information about another. In the case of the particle and the detector in the double-slit experiment, for example, let |left⟩ and |right⟩ be the states (waves) in which the particle goes through either the left or the right slit, respectively, and let |click⟩ and |no click⟩ be the states (waves) in which the detector behind the right-hand slit clicks or does not click. Let |ready⟩ be the state of the detector before the measurement takes place, in which it is ready to detect the particle if the particle goes through the right slit. Just before the measurement, the particle is in the superposition state |left⟩ + |right⟩ and the detector is in the state |ready⟩. During the measurement, in the |left⟩ part of the superposition the particle goes through the left slit and the detector does not click, while in the |right⟩

part of the superposition the particle goes through the right slit and the detector clicks. Just after the measurement, then, the state of the particle and detector is the superposition |left, no click⟩ + |right, click⟩. That is, particle and detector are in an entangled state that is a superposition of particle-through-left-slit correlated with no-click-from-detector and particle-through-right-slit correlated with click-from-detector.

Suppose I am in the room while the double-slit experiment is performed, and either I hear a click or I don't. I, too, am a quantum system, albeit one composed of many pieces. Let |Seth hears click⟩ be the wave that corresponds to my hearing a click and |Seth doesn't hear click⟩ be the wave that corresponds to my not hearing a click. (Note that these waves are quite complicated waves, corresponding to all the atoms in my body.)

After the sound (if any) has reached my ear, the state of the system comprising the particle, the detector, and me is |left, no click, Seth doesn't hear click⟩ + |right, click, Seth hears click⟩. I have become entangled with the particle and the detector. In this entangled state, you can see that my state relative to the particle being on the right and the detector clicking is |Seth hears click⟩. My state relative to the particle being on the left and the detector not clicking is |Seth doesn't hear click⟩. The quantum tree fell in the quantum forest, and there was a quantum someone there to hear it.

This "relative state" picture of the quantum-measurement process illuminates the phenomenon of measurement. The information about which slit the particle went through infects first the detector, then me. If I write you a letter telling you whether or not I heard a click, then when you receive the letter your relative state will reflect what happened: |Seth wrote me to say he heard a click⟩ or |Seth wrote me to say he didn't hear a click⟩. Now *you* have become entangled with the particle, the detector, and me. After the measurement, the information about its outcome spreads to infect whatever it comes into contact with.

Despite the fact that the relative-state picture of quantum measurement illuminates the phenomenon, there is something disquieting about it. When I hear a click, what happens to the other part of the superposi-

Figure 10. Schrödinger's Cat

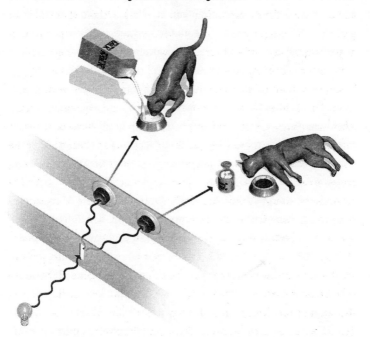

Two buttons are substituted for the two slits in the double-slit experiment. If the first button is pressed a cat is fed milk; if the second is pressed, the same cat is fed poison instead. But the quantum-mechanical particle presses both buttons: Schrödinger's cat is both alive and dead at the same time!

tion, in which I didn't hear the click? Is the person who didn't hear a click still me? Did I both hear and not hear a click at the same time? The unnerving aspect of this picture was amplified in 1935 by the Austrian physicist Erwin Schrödinger, who imagined that when the detector clicks, it trips a mechanism that kills a cat. In the Schrödinger's Cat paradox, the state of particle, detector, and cat after the measurement is |left, no click, cat alive⟩ + |right, click, cat dead⟩. In some weird quantum way, the cat is alive and dead at the same time.

The Schrödinger's Cat paradox has inspired much confusion. Stephen Hawking got so tired of this paradox that he would often say (in paraphrase of Hermann Goering and Hanns Johst), "Whenever I hear the words 'Schrödinger's cat,' I reach for my gun." The original solution to the Schrödinger's Cat paradox, advocated by Bohr, was that when you hear the click and the cat dies, the other part of the superposition—the one in which you don't hear the click and the cat lives—simply goes away. This disappearance of the parts of the wave that we don't experience is an instance of the wave-function collapse explained earlier, since the wave in effect collapses into one of its components. In the wave-function-collapse picture of quantum measurement, by the time I write to tell you that I heard a click and when I turned around the cat was dead, the part of the wave in which the cat is still alive has disappeared.

The problem with this solution, as we have seen, is that the dynamical laws of quantum mechanics are reversible. In principle, it is possible to return to the original state, the one before the measurement was made. If the wave function actually collapses, it isn't possible to perform this reversal; however, in many cases, as in the spin-echo effect and its analogs, you can reverse the dynamical evolution of a many-piece quantum system and return it to its original state. Theory and experiment make wave-function collapse an implausible solution to the measurement problem.

Fortunately, there is a simple and elegant alternative to the wave-function collapse explanation. *The measurement problem stems from the presence of those parts of the wave function corresponding to alternatives that do not actually happen.* It would be nice to be able to ignore them. That is, when the detector has clicked and I have written to tell you the cat is dead, I would like to put the matter to rest and cease to worry about the parts of the wave function in which the cat is still alive. Let bygones be bygones. When are we justified in ignoring these other parts of the wave function? The answer to this question was supplied by Robert Griffiths and Roland Omnes and further elaborated by Murray Gell-Mann and James Hartle: *We can ignore the other parts of the wave function at exactly the moment when they have no further effect on us.*

This solution to the measurement problem depends not only on the present but on the future. If the other parts of the wave function will never again interfere with ours, then we say that the future history of the wave function *decoheres*. This "decoherent-histories" approach to quantum mechanics neatly resolves most of the troubling aspects of the measurement problem.

In the case of the double-slit experiment, for example, there are two possible histories. In one of them, the particle goes through the left slit and lands on the wall. In the other, the particle goes through the right slit and lands on the wall. These histories are coherent, not decoherent: they interfere with each other to create the pattern of bands on the wall.

Now add the detector to the right-hand slit. There are still two possible histories. In one of them, the particle goes through the left slit and lands on the wall. In the other, the particle goes through the right slit, trips the detector, and lands on the wall. Because of the detector, the interference pattern goes away. These histories are decoherent: they do not interfere with each other. Similarly, in the Schrödinger's Cat paradox, once the detector has clicked and the cat is dead, looking at the cat again to see if it is still dead makes no difference for the future: the cat stays dead. The histories of that experiment are thus decoherent. In this case, we can say that the cat is either dead or alive, but not both.

There is a simple criterion for deciding whether a set of histories is coherent or decoherent. Think of what happens when you make a measurement. Measurement destroys coherence. But it can't destroy coherence if there is no coherence there to destroy. If making a sequence of measurements on a quantum system changes its future behavior, then the histories corresponding to the possible sequences of outcomes of the measurements are coherent. If the sequence of measurements has no effect on the system's future behavior, the histories are decoherent. In the double-slit experiment, measurement destroys the interference pattern and changes the behavior of the system: The histories of the double-slit experiment are thus coherent.

Many Worlds

The decoherent-histories picture of quantum mechanics provides an intuitively satisfying resolution to the measurement problem. During measurement, the particle and detector become entangled and the wave function is a superposition of two states. One of these states corresponds to what "actually happens." As long as the future history of particle and detector (and the cat and me and you) is decoherent, then the other state has no further effect.

The other state—the other part of the wave function—is in some sense still there, however, even though we can safely ignore it. This feature has led some people to advocate the so-called Many Worlds interpretation of quantum mechanics, according to which this other part of the wave function corresponds to another world, in which the cat is happily alive. The cat, say Many Worlds advocates, is really dead and really alive at the same time.

There is some division in the physics community about the Many Worlds interpretation of quantum mechanics. In 1997, I debated the issue with the Oxford physicist David Deutsch, who is a strong advocate of the Many Worlds picture. I'm not sure who won the debate in this particular world.* For the rest of this book, I will use the "many histories" interpretation of quantum mechanics advocated by Gell-Mann and Hartle. In this interpretation, quantum mechanics supports sets of decoherent histories as described above. Out of this set, one history really happens. The remainder of the histories correspond to inaccessible pieces of the wave function. These histories correspond to possible events that

*For the full text of the debate, see http://www-me.mit.edu/people/personal/slloyd.htm.

didn't really happen. (Or, as Deutsch would urge us to say, that didn't really happen in this world.)

In my opinion, the Many Worlds picture does injustice to the word "really." Normally, people use the word "really" to refer to things that are actually the case: I really wrote these words; you are really reading them. There are other parts of the wave function in which I wrote something else and you are watching TV. But those parts of the wave function don't correspond to what *really* happened. They are like the forking paths in Borges's story: even if they are there, they have no effect on reality.

CHAPTER 6

Atoms at Work

Talking to Atoms

New York City is full of people wandering the streets talking to the empty air. When asked what they are doing, they claim the air is talking back in voices only they can hear. One morning, when I was a graduate student in New York, I was having breakfast at the counter of the Polish coffee shop just around the corner from my apartment on Second Street at Avenue B. As I dug into my kielbasa and eggs, the man sitting next to me grabbed my arm, gazed into my eyes, and said, "They took Einstein's brain and transplanted it into my head."

"Really?" I said. "Then there's some questions I need to ask you." I proceeded to ask him his opinions on quantum mechanics and general relativity. Unfortunately, in light of his answers, the transplant seems to have been less than successful.

Einstein was somewhat confused about quantum mechanics. As noted, he never fully believed or trusted in the theory, objecting to its intrinsically chancy nature. Quantum mechanics went against his powerful intuition, just as it goes against pretty much everyone else's, and Einstein had more right than most to trust his instincts. But Einstein's intuition here led him astray. Quantum mechanics *is* inherently chancy, and there are literally millions of experiments that confirm the accuracy of the theory.

To see how quantum mechanics injects chance into the universe, it's useful to contemplate a simple example of a throw of the quantum dice (in my job as an atomic masseur, I have some practice in the quantum crap game): Take an atom. Zap it with a laser. Now zap the atom with another laser and look to see if it emits light. If it emits light, call that 0. If it doesn't emit light, call that 1. Half the time, the atom will emit light (a 0), and half the time it will remain dark (a 1). Thus, a brand-new bit is born.

Let's look more closely at atom zapping, a process that allows us to talk to atoms, and to hear them talk back. Unlike my conversation with the New York street person, a real answer emerges: the atom replies to your question by emitting light, or not. To understand the significance of the atom's answer, we must learn some of the language of atoms.

You are to an atom as Earth is to an ant: very large. Atoms are typically a few ten-billionths of a meter across—tiny, bouncy spheres held together by electricity. An atom consists of a compact nucleus (Latin for "nut") 100,000 times smaller still, made up of protons (which are positively charged) and neutrons (lacking a charge). Most of the mass of the atom lies in its nucleus, which is surrounded by a cloud of electrons, whose masses are a couple of thousand times smaller than those of protons or neutrons. Electrons are negatively charged, so they are attracted to the positively charged nucleus; there are the same number of electrons in the surrounding cloud as there are protons in the core, so the atom as a whole is electrically neutral.

The electric force binds electrons to the nucleus. When an atom is in its normal state, called the "ground state," the electrons are clustered as close as they can be to the nucleus. (The nucleus itself is bound together by the so-called strong force, which is 1,000 times stronger than the electromagnetic force.) What does "as close as they can be" mean? Why don't the electrons just fall into the nucleus, bonking the atom on its nut? In fact, classical mechanics predicts just this bonking process. If classical mechanics were correct, atoms would survive for only a tiny fraction of a second before disintegrating in a burst of light. But the correct picture of atoms is given by quantum mechanics, not classical mechanics. Quantum mechanics guarantees the stability of atoms, and the stability of atoms, in turn, is one of the most concrete confirmations of quantum mechanics. Without quantum mechanics, the life of an atom would be exciting but short.

But how does quantum mechanics guarantee the stability of atoms? Recall that each electron has a wave associated with its position and velocity. The places where an electron's wave is big are where the electron is likely to be found. The shorter the length of the wave, the faster the electron is moving. Finally, the rate at which the wave wiggles up and down is proportional to the electron's energy.

Suppose we want to fit an electron's wave around an atom's nucleus. The simplest wave that can fit around a nucleus is a sphere: the wave wraps smoothly all the way around. The next simplest wave has one peak as it wraps; then comes a wave with two peaks, and so on. Each of these types of waves corresponds to an electron in a definite energy state. The simplest wave is the spherical one with no peaks; in this state, the ground state, the electron has the lowest energy. The second wave wiggles; the electron has more energy. The more peaks in an electron's wave, the faster it wiggles and the greater its energy.

Attach a rock to a rubber band and whirl it around your head. The faster the rock moves, the more energy it has, and the farther away from your head it whirls, because the rubber band has to stretch to compensate for the additional speed of the rock. The same holds for the electron: the more energy it has, the farther away from the nucleus it orbits. The electron can snuggle up closest to the nucleus when it has its minimum energy, which occurs when it is in the simple, spherical wave, or it can orbit farther away. Wave-particle duality implies that an atom's electrons consist of a set of discrete waves, so there are only so many orbits they can take. They never fall into the nucleus, and we can count the possible options (no peaks, one peak, two peaks, and so on).

When an electron jumps from a higher energy state to a lower one, it emits a chunk, or quantum, of light—a photon—whose energy is equal to the difference between the energies of the two states. Atoms of different kinds—a phosphorus atom, for example, with fifteen electrons, or an iron atom, with twenty-six—emit photons with characteristic energies. Because of the correspondence between energy and the rate at which the emitted photons wiggle up and down, these photons correspond to light of a characteristic frequency. These frequencies are called an atom's "spectrum."

The fact that atoms emit light with characteristic spectra was observed in the first half of the nineteenth century. Because they did not know about quanta or photons, classical physicists were unable to account for these spectra. The explanation of atomic spectra was the first great triumph of quantum mechanics. By using the simple relationships between a wavelength and the speed of electrons, and between the frequency with which a wave wiggles and its energy, Niels Bohr was able to calculate the spectrum of the hydrogen atom, showing that this quantum-mechanical model agreed closely with findings observed by experiment.

Not only can atoms emit light, they can also absorb it. Just as an atom can jump from a higher energy state to a lower one, emitting a photon in the process, an atom can absorb a photon and jump from a lower energy state to a higher one. Take an atom in its ground state and bathe it with a beam of laser light made up of photons whose energy is equal to the difference in energy between the ground state and the next-lowest energy state (called the first "excited state") of the atom. The atom will absorb a photon from the beam and jump from the ground state to the first excited state.

If the atom is bathed by photons whose energy is not equal to the energy difference between the state it's in and some higher energy state, then it will not absorb the photons. Atoms can absorb energy only in specific chunks (quanta). This feature is useful for controlling the state of atoms. If the atom is bathed with photons of the wrong energy, it refuses to absorb any photons and will just stay where it is, whereas if an atom is bathed with photons whose energy equals the difference in energy between its current state and some higher state, the atom will happily absorb a photon and jump to that state. The fact that atoms respond to light only at frequencies corresponding to their spectrum is useful if you want to send instructions to one kind of atom but not to another, as we will see.

Going from state to state by emitting or absorbing a photon takes a characteristic amount of time that depends on the intensity of the laser beam. In particular, it is possible to subject an atom to a pulse of laser light with the following result: if the atom is in its ground state, it jumps to the first excited state, absorbing a photon in the process; and if the atom is in its first excited state, it jumps to the ground state, emitting a photon in the process. The ground and first excited state of an atom correspond to a bit. We can take the ground state to correspond to 0 and the first excited state to correspond to 1. But the atom is not just a bit; it is a qubit. The atom's states correspond to waves, just like the states of the nuclear spins described earlier. So in keeping with our practice of bracketing quantum-mechanical things, we call the ground state $|0\rangle$ and the first excited state $|1\rangle$. Applying a laser pulse to the atom takes $|0\rangle$ to $|1\rangle$ and takes $|1\rangle$ to $|0\rangle$. In the language of atoms, the atom is simply going from state to state; in the language of zeros and ones, this is the famous logic operation called NOT. By speaking the language of atoms, we can make an atom flip its bit.

How do we get the atom to talk back to us? Just as we can address the atom using light, the atom can talk back to us using light. Imagine a third state, $|2\rangle$, higher in energy than the qubit states $|0\rangle$ and $|1\rangle$. Suppose that whenever the atom is in state $|2\rangle$, it tends to jump back to $|0\rangle$, the ground state, spontaneously emitting a photon in the process. Spontaneous emission of photons is responsible for the phenomenon known as fluorescence. A fluorescent light works by exciting atoms out of their ground state and letting them jump back, emitting light in the process. The energy of the emitted photon is equal to the difference in energy between the state $|2\rangle$ and the state $|0\rangle$. If you look closely—through a microscope, for instance—you can sometimes see the emitted photon as a flash of light. The atom is talking to you.

The ability to see spontaneously emitted photons allows us to determine whether or not the atom is in the ground state. Bathe the atom with

Figure 11. Flipping Qubits

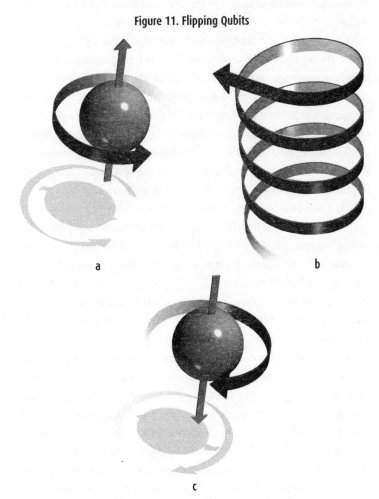

a

b

c

To flip a quantum bit, simply shine light on it. Figure 11a shows a qubit—a nuclear spin—in the state spin-up, or 0. In figure 11b a particle of light, or photon, arrives. It is absorbed by the nuclear spin, flipping it to the state spin-down, or 1, in the process (figure 11c).

light made up of photons each of whose energy is equal to the energy difference between the states $|0\rangle$ and $|2\rangle$. If the atom is in the ground state, then because the photons in which it is bathed have just the right energy, it will absorb a photon and go to state $|2\rangle$. Shortly afterward, it will emit a photon and jump back to the ground state. Then it will absorb another photon and jump back up to state $|2\rangle$. Then it will emit a photon and jump back to the ground state. Such a process, in which an atom keeps on absorbing and emitting photons, is called a "cycling transition," because the atom cycles back and forth between two well-defined states.

If the atom starts out in the state $|1\rangle$, by contrast, then it can't absorb a photon and go to the state $|2\rangle$, because the photons available have the wrong energy. An atom starting in the state $|1\rangle$ will simply stay there, impervious to the bath of photons, not fluorescing. An atom that fluoresces is saying, in effect, "I'm 0! I'm 0! I'm 0! I'm 0!"

Now look more closely at how atoms jump from state to state when zapped by a laser. Start an atom off in its ground state and bathe it in light made of photons whose energy is equal to the difference in energy between the ground and the first excited states. What happens during the jump? During the jump, the atom and the light are in a state that is a *superposition* of the ground state of the atom, with no photon absorbed from the bath, and the first excited state of the atom, with one photon absorbed. That is, the state of the atom is a superposition of two waves. The first wave is the ground state and the second wave is the first excited state. Just after the atom begins its bath and starts its jump, the superposition is made up mostly of the ground state, with only a little bit of the excited state mixed in. Halfway through the jump, the atom and bath are in the equal superposition state $|0$, no photon absorbed$\rangle + |1$, photon absorbed$\rangle$. Near the end of the jump, the superposition is mostly the excited state, with only a little bit of the ground state remaining.

An atom does not instantaneously jump from its ground state to its excited state. Rather, it glides through a continuous intermediate sequence of superpositions. The same continuous gliding occurs when an atom goes from the first excited state back down to the ground state, emitting a photon in the process. The atom and the bath of photons start out in the state |1, no photon emitted⟩ and end up in the state |0, photon emitted⟩. Halfway through the jump, the atom and the bath are in the superposition state |1, no photon emitted⟩ + |0, photon emitted⟩.

This description of an atom rotating from one state to another while absorbing and emitting photons is reminiscent of the earlier description of a nuclear spin rotating from one state to another while interacting with a magnetic field. And in fact, these two processes are essentially the same. As a nuclear spin rotates, it, too, absorbs a photon—from the magnetic field—and emits a photon as it returns to its original state.

Now you know how to talk to atoms. By zapping an atom with a laser, you can control its state. You can continuously rotate the atom through a sequence of superposition states; you can excite it, causing it to absorb a photon, and de-excite it, causing it to emit a photon. You also know how to make the atom talk back. By driving cycling transitions, you can ask an atom whether it registers 0 or 1, and you can get a response. All of this means that you now have the power to create new bits.

Take an atom and zap it with a laser to put it in the superposition state |0⟩ + |1⟩. Now drive a cycling transition to see if it is in the state 0 or the state 1. If it is 0, the atom will fluoresce; if it is 1, it will remain dark. You have tossed the quantum coin to create a brand-new bit.

Talking to an atom by driving a cycling transition measures the state of the atom and creates information. Of course, just as in the last chapter, what happens during the measurement is open to interpretation. In the wave-function-collapse interpretation of measurement, the wave function of the atom taken together with the photons has collapsed to either the state |0, fluorescence⟩ or to the state |1, no fluorescence⟩.

In the decoherent-histories interpretation, the state of the atom together with the photons is in the superposition state |0, fluorescence⟩ + |1, no fluorescence⟩. Each of the states in the superposition corresponds to a decoherent history. In this case, the histories are *highly* decoherent. To make them cohere would require that you collect all the photons emitted by the atom, reflect them back, and force the atom to reabsorb them. You would need a kind of Loschmidt's demon, capable of reversing the sequence of events in time. But reversing photons that are scattering all over the universe is hard (if you doubt it, then go ahead, reverse them). So, because the two states in the superposition decohere, the atom and photon behave as though they are in one state or the other, and you do indeed generate a brand-new bit, one that never existed before.

Quantum Computation

If you zap an atom with light whose photons have the right energy, you can make the atom flip its state from $|0\rangle$ to $|1\rangle$ and back again. You are flipping the atom's bit. In other words, you are performing the logical operation known as NOT. In a 1993 article in *Science* entitled "A Potentially Realizable Quantum Computer," I showed how a slightly more involved set of laser pulse sequences than those used to perform NOT operations allows atoms to perform the logic operations AND, OR, and COPY, as in conventional digital computations. Each atom stores one bit, and a collection of atoms can compute anything an ordinary PC or Macintosh can compute.

But they can do much more than an ordinary PC or Macintosh. Atoms register more than bits; they register qubits. Unlike classical bits, qubits can be in quantum superpositions of $|0\rangle$ and $|1\rangle$; that is, they can register 0 and 1 at the same time. *Is there any way these quantum superpositions can be used to compute in ways that classical computers cannot?* This question was first raised by David Deutsch in the mid-1980s, but it was not until the early 1990s that the question was definitively answered. *The answer is Yes.*

To see why quantum computers and quantum bits can do more than classical computers and classical bits, think about what bits do in a computer. Some bits, like those in the computer's hard drive or memory, just store information. For example, bits in my computer's memory receive and store this text as I type it. Other bits, like those in computer programs, are instructions or commands. They tell the computer to do one thing or another. Whether a bit functions as a memory bit or a command bit depends on the context in which it is used.

Consider a bit the computer interprets as a command: 0 means "Do this!" and 1 means "Do that!" "This" could mean, say, "Add 2 plus 2" and "that" could mean "Add 3 plus 1." Or "this" could mean "Send an e-mail" and "that" could mean "Fire up the Web browser."

Unlike a classical bit, a quantum bit can register 0 and 1 at the same time. What does a quantum computer do when it tries to interpret such a qubit as a command? The 0 part of the superposition is telling the quantum computer to "Do this" while the 1 part of the superposition is telling the quantum computer to "Do that." How does the quantum computer decide? It doesn't. Instead, it does "this" and "that" at the same time! Just as a quantum bit can register two values at once, *a quantum computer can perform two computations simultaneously.*

David Deutsch called this strange ability of a quantum computer to do two things at once "quantum parallelism." Quantum parallelism is quite different from ordinary classical parallel computation. A classical parallel computer consists of several processors linked together. In a classical parallel computation, one processor performs one task while the other processors perform other tasks. In quantum parallelism, a single quantum processor performs several tasks at once.

This ability to do two things at once is intrinsic to quantum mechanics. The photon in the double-slit experiment can go through both slits at once; a qubit can register 0 and 1 at the same time; and a quantum computer can perform two distinct tasks simultaneously. The ability to do two things at once arises from the wave nature of quantum mechanics. Each possible state of a quantum system corresponds to a wave, and waves can be superposed.

We're all familiar with situations in which superposing waves results in qualitatively new and richer phenomena. Consider sound waves. A wave wiggling up and down at a particular frequency corresponds to a pure tone. A sound wave wiggling up and down 440 times per second is the sound of the note A above middle C. A sound wave wiggling up and down 330 times per second is the sound of the note E above middle C. The superposition of these two waves corresponds to a chord, which is qualitatively different from and richer than the sound of either pure tone taken by itself. The richness of the sound arises from the way in which the two pure notes interfere with each other.

A classical computation is like a solo voice—one line of pure tones succeeding each other. A quantum computation is like a symphony— many lines of tones interfering with one another. This interference phenomenon is what gives quantum computation its special qualities and added power.

Quantum computations are not restricted to just two "voices." Like a symphony, a quantum computation gains its power by building up intricate sequences of chords. For example, suppose the computer is given as input a "qutrit," with three possible states, 0, 1, and 2. The 0 state instructs the quantum computer to "do this," the 1 state instructs it to "do that," and the 2 state instructs it to "do the other thing." In the example in which "this" means "Add 2 plus 2" and "that" means "Add 3 plus 1," "the other thing" might be "Add 4 plus 0." When the quantum computer is given a superposition of all three instructions as input, it does "this," "that," and "the other thing" all at once. In our example, the computer is simultaneously exploring all ways of constructing 4 as the sum of nonnegative integers. Such a quantum computation is like a trio, in which three waves interfere with each other and three computational "voices" cooperate to deconstruct the number 4 faster than one computational voice could on its own.

The number of things a quantum computer can do at once—the number of voices in the symphony of quantum computation—grows rapidly as the number of bits of input increases. Even a small number of qubits allow an extraordinarily rich texture of interfering waves as they compute. A quantum computer given 10 input qubits can do 1,024 things at once. A quantum computer given 20 qubits can do 1,048,576 things at once. One with 300 qubits of input can do more things at once than there are elementary particles in the universe. Quantum parallelism allows even a relatively small quantum computer, containing only a few hundred qubits, to explore a vast number of possibilities simultaneously.

The Measurement Problem, Again

What happens when you take a quantum computer that is doing several things at once and ask it what it's doing? Is it possible to take a measurement to determine if it is doing this, that, or the other thing? As with any quantum system, when you take a measurement in a superposition of several possible states, the result of the measurement yields one of those possibilities at random. So if the quantum computer is exploring all the ways of constructing 4 as the sum of positive integers, when you measure it, it will tell you, for example, "Oh, I was adding 3 plus 1," or "I was adding 2 plus 2."

To pursue the metaphor of quantum computation as symphony: If you measure the quantum computer while it is computing, you don't hear the full effect of the orchestration; rather, you hear one of the voices selected at random.

Recall the double-slit experiment. In that model, the electron does two things at once: it goes through both slits simultaneously. When you take a measurement to determine which slit the electron has gone through, it will show up at one slit or the other at random. Similarly, when you take a quantum computer that is doing two things at once and measure to see what it's doing, you will find it doing one or the other of those things at random. If you want to see the interference pattern in the double-slit experiment, you must wait until the electron has hit the screen, so that the two waves—one from one slit, one from the other— can interfere with each other. The interference pattern comes from the "duet" of the two waves. In a quantum computation, if you wish to get the full benefit of the computation, you must not look at the computation while it is occurring. To get the full symphonic effect of a quantum computation, you must let the all the waves in the computation interfere with one another. You must let the "voices" of the computation blend together on their own.

One way of looking at this phenomenon is to say that measuring a quantum computer that is doing several things at once "collapses the computer's wave function," so that it ends up doing just one thing. Another way of describing the effect of such a measurement, though, is to say that it "decoheres the computation." As discussed earlier, decoherence does not suppose that the alternate possibilities have entirely gone away, rather that they still exist, but no longer affect the state of the system as we know it.

Note that a full-blown measurement is not necessary to decohere a quantum computation. Any passing electron or atom that interacts with the quantum computer in such a way as to get information about what the quantum computer is doing can decohere the computer as effectively as a full-blown measurement using a macroscopic measuring device. Great care must be taken to insulate quantum computers from their surroundings while they are performing quantum computations.

Factoring

Quantum parallelism makes quantum computers potentially very powerful. A quantum computer can explore all possible solutions to a hard problem at the same time. An example of such a problem is factoring. A number is factorizable if it can be written as the product of two or more integers greater than 1. For example, 15 is factorizable because it can be written as 3 times 5. But 7 is not factorizable, because the only way it can be written as the product of two positive integers is 7 times 1. Numbers that are not factorizable are called prime numbers, or primes. The first few prime numbers are 2, 3, 5, 7, 11, 13 . . . It's not hard to show that there are an infinite number of primes.

Take two large prime numbers, each with 200 digits, and multiply them. The result is a 400-digit number. Multiplying two 200-digit numbers is tedious, but it's a straightforward task for a digital computer, classical or quantum. Take the resulting 400-digit number, hand it to someone who doesn't know the original two prime numbers, and ask him to factor it. The 400-digit number is clearly factorizable, and if you know the two 200-digit factors, it is straightforward to verify that they can be multiplied together to give the proper 400-digit result. But finding those two factors if you don't know them beforehand proves daunting. In fact, the only known way to find those factors is, essentially, to examine all possible 200-digit numbers in turn until you find one that divides the 400-digit number. (There are tricks allowing you to eliminate some numbers from consideration, but they don't help that much.) Unfortunately, there are a lot of 200-digit numbers. To use our favorite large number, there are more 200-digit numbers than there are elementary particles in the universe.

There is no known easy way to use a classical computer to factor a 400-digit number. One of the largest classical computations ever performed was factoring a 128-digit number a few years ago. The computation used hundreds of classical computers linked by the Internet and required trillions of logical operations to be performed on billions of bits. More recently, a 200-digit number was factored. Factoring a 400-digit number using conventional techniques is likely to remain out of reach for many years.

The apparent difficulty of factoring large numbers is the basis for a powerful method for protecting information. Whenever you use your bank card or buy anything over the Internet, the security of your transaction is protected by a technique called public key cryptography. Suppose you wish to use your credit card to buy extra copies of this book from Amazon.com. Amazon sends you a "public key" consisting of a large number that is the product of two smaller prime numbers. Your computer then uses this public key to scramble or "encode" the information you send to Amazon, including your credit card information. To decode that information, Amazon uses the "private key," consisting of the two prime numbers that, multiplied, give the public key. Thus anyone who possesses the public key can encode information, but to decode it requires the private key, consisting of its factors. Public key cryptography is clearly a useful trick, and its security depends on factoring being hard. Public keys of 256 digits are very hard to crack by classical computation and are currently considered more than adequate to protect most forms of information.

In 1994, however, Peter Shor at AT&T Laboratories showed that even a relatively small quantum computer, with only a few thousand qubits, could factor 400-digit numbers with ease. In effect, he orchestrated the computation so that the actual factors could be identified over the noisy background of potential factors. To see how factors can be identified by combining their waves in a quantum computation, again think of a symphony: if Beethoven orchestrates his theme to be played by violin, cello, piccolo, and trombone, you will hear that theme no matter what the rest of the orchestra is playing.

Suppose that while the quantum computer is exploring all possible factors, you rudely go in and measure the quantum computer's qubits to find out what the computer is doing. Its answer will be "Oh, I was just looking at [some pair of 200-digit numbers] to see if, when multiplied, they gave the right answer." Most of the time, the numbers don't. Measuring a quantum computer while it's exploring all possible solutions to the factoring problem is no different from picking one of those possible solutions at random. To get the full benefit of the computation, you must not disturb the computer while it's computing. You must let each of the parallel computations take its course, interfering with the others; only then can the symphonic nature of quantum computation help you find the factors.

Searching

Factoring is not the only hard problem that quantum computers are potentially good at solving. In 1996, Lov Grover, of Bell Laboratories, showed that quantum computers were better at searching than classical computers. Suppose you don't remember in which of your four pockets you put your wallet. First you check one pocket, then another. In the worst case, you need to root around in all four pockets to find your wallet; on average, you need to check two. But suppose you could use quantum parallelism to check all your pockets at once. Grover showed that you need make only one quantum search to come up with the wallet.

Needless to say, Grover's algorithm works with more than four options. If you're looking for something that could be in 100 possible places, you need to perform only 10 quantum searches to find it, compared with an average of 50 classical searches. If you're looking for something in a million possible spots, you need to perform only 1,000 quantum searches, compared with half a million classical ones. In general, the number of quantum searches required to locate what you're looking for is the square root of the number of places in which it could be.

What other problems can quantum computers solve more efficiently than classical computers? In order to exploit the symphonic nature of quantum parallelism, you must allow all the parts of the quantum computation to interfere with one another. But like writing a symphony, arranging the necessary quantum interference is tricky, so much so that there are only a few quantum algorithms, such as factoring and searching, that are currently better than their classical analogs.

Building Quantum Computers

In the fall of 1994, I received a call from Jeff Kimble, a professor of physics at the California Institute of Technology. Jeff had read a couple of my papers on quantum computation and wanted to discuss the possibility of constructing quantum logic gates using photons.

Jeff Kimble is a tall Texan who has a way with atoms and light. He was introduced to me as the man who had "squeezed light more than it had ever been squoze before," and if that light felt anything like my hand after he had shaken it, I was inclined to believe this description. When we began talking, I noticed two things. The first was that Jeff tells you without hesitation what he thinks. If he looks at your calculation and says in his soft Texas accent, "There's trouble in River City," that means you're in trouble. The second was that as he described his experiments, I didn't understand more than one word in three of what he was saying.

Week by week that fall, as I spent more time talking with Jeff and his students, the picture became clearer. He was taking individual photons and making them interact strongly with individual atoms. Basically, he was putting a photon in a can with an atom and shaking them about. The "can" was an optical cavity consisting of two mirrors a few millimeters apart. The photon bounced back and forth between the mirrors tens of thousands of times before eventually getting free. Jeff would drip cesium atoms into the cavity while sending in the photons from his lasers and then see what came out. Because each photon and each atom spent a substantial fraction of a second confined together in a small space, they had plenty of time to interact.

Whereas an atom is tiny, one ten-billionth of a meter across, a photon can be a whole lot bigger. Recall that photons are particles of light, arising out of wiggles in the electromagnetic field. Because of the Heisenberg uncertainty principle, there is a trade-off between the rate at which the electromagnetic field is wiggling and the volume the photon occupies: the better defined the field's wiggle rate, the larger the volume of space the photon occupies. In Jeff Kimble's optical cavities, the photons that can get in have a very well defined frequency and are long and skinny— 100 meters long! But the cavity itself is only a few millimeters in length. How can something 100 meters long fit into a container so much smaller? It took me some time to bend my mind around this question. The answer is that the photon goes into the cavity the way a snake goes into a coffee can: It gets doubled back and forth thousands of times. Because of all of those doublings, the strength of the electromagnetic field that corresponds to a single photon is thousands of times higher inside the cavity than outside. As a result, a photon in the cavity interacts *very* strongly with an atom in the cavity.

Jeff and his graduate students Quentin Turchette, Christina Hood, and Hideo Mabuchi were then performing experiments in which they sent two photons through the optical cavity in the presence of an atom and looked at what had happened to the photons when they came out. Because both photons interacted strongly with the atom, they interacted strongly with each other as well. But was that interaction enough to construct a quantum logic gate, such as a controlled-NOT gate? When we started, they didn't know from quantum logic gates, and I didn't know from photons. When we were done, I had shown that almost any interaction between photons sufficed to construct a quantum logic gate, and they had exhibited the first photonic quantum logic gate.

At the same time, Dave Wineland and Chris Monroe at the National Institute for Standards and Technology (NIST) in Boulder performed an experimental realization of a proposal by Ignacio Cirac and Peter Zoller of the University of Innsbruck for an architecture for quantum computers, based on trapping ions using oscillating electromagnetic fields and then zapping them with lasers. These ions were atoms that had been stripped of an electron and thus had a net positive charge, allowing them to be trapped with relative ease. (Old physics joke: Two atoms walk into a bar. One atom looks down at himself and says, "Hey, I lost an electron." The other atom says, "Are you sure?" The first atom replies, "I'm positive!") Once the ions were trapped, they could be cooled to very low temperatures and zapped with lasers to make them interact and perform quantum logic operations.

Both Kimble's and Wineland's experiments were performed by modifying existing experimental setups. The relative ease with which the first quantum logic operations were performed made it plausible that quantum computers could in fact be constructed.

When I joined the faculty of MIT in December 1994, I began working with scientists and engineers from all over the world to build quantum computers. David Cory, a professor of nuclear engineering at MIT, with his colleagues Tim Havel and Amir Fami, had shown how nuclear spins could be made to compute using the atom-zapping techniques described above; when applied to nuclear spins, these atom-zapping techniques are called "nuclear magnetic resonance," or NMR. Shortly afterward, Neil

Gershenfeld at the MIT Media Lab independently discovered how to use NMR to compute, and I began working with Neil and Isaac Chuang to perform simple quantum computations using NMR. With only a two-bit quantum computer, we were able to provide demonstrations of quantum parallelism by making it perform several tasks simultaneously. Later, Chuang would build larger and larger NMR quantum computers, culminating, a few years ago, in a seven-qubit computer that could perform a simple version of Shor's algorithm. Chuang used this computer to factor the number 15. Clearly, there is still a long way to go before we have quantum computers that can factor 400-digit numbers!

Figure 12. A Simple Quantum Computer

A simple quantum computer can be constructed out of a chain of nuclear spins in a molecule. Here the nuclear spins are spin up, up, down, down, up: that is, the spins register the quantum bits 00110.

In addition to continuing to collaborate with Jeff Kimble on storing and transporting quantum bits on photons, I began working with scientists at MIT on making light interact with atoms. Jeffrey Shapiro, Franco N. C. Wong, and Selim Shahriar, all of the Research Laboratory of Electronics, were keen to explore the possibilities for quantum communication, and we soon wrote a proposal for constructing the world's bright-est source of entangled photons, together with a method for catching those photons using atoms trapped in optical cavities. These techniques, along with related techniques proposed by Kimble, Cirac, and Zoller, form the basis for an attempt to construct a quantum internet—a net-work of quantum computers linked together optically. (I am currently working to design a quantum Internet search engine, provisionally called "Quoogle.")

At the same time, I began to collaborate with Hans Mooij, of the Delft University of Technology, on the possibility of constructing quantum computers using superconducting systems. In a superconductor, elec-trons encounter almost no resistance as they move from place to place. Such a flow of superconducting electrons is called a supercurrent. Another way of thinking about superconductivity is that the atoms in the materials through which the electrons move have a hard time grabbing hold of the electrons. This means that the electrons can move through the material in a way that preserves quantum coherence: they stay entan-gled. Mooij and other workers on superconductivity had pointed out that it might be possible to use this slippery feature of superconducting electrons to perform quantum computation. If you construct a super-conducting loop of material and interrupt it in the right places with very thin pieces of non-superconducting material called Josephson junctions, the resulting device can support supercurrents that go either clockwise or counterclockwise around the loop. Such a device could easily store one bit of information: identify counterclockwise supercurrents with the state 0 and clockwise supercurrents with the state 1.

Such a superconducting quantum system can register not just a bit but a qubit. We calculated that if we designed the superconducting bit very carefully, in order to reduce interactions between the supercurrent and its surroundings to the absolute minimum, then the supercurrent could be put in a quantum-mechanical superposition of circulating clockwise and counterclockwise at once. At some level, the ability of a supercurrent to exhibit such a quantum superposition should not be surprising; after all, a supercurrent consists of electrons, and an individual electron has no trouble being in two places at once. But a supercurrent can consist of billions of electrons, and the loop around which it circulates clockwise and counterclockwise at the same time is almost large enough to be seen with the naked eye. Such macroscopic quantum coherence is indeed surprising, and researchers had been trying for decades to exhibit it without success. With Terry Orlando at MIT, Mooij and I began a research collaboration that resulted a few years later in the demonstration by Mooij's student Caspar van der Wal of quantum bits that could be put in just such a macroscopic quantum superposition. (James Lukens's group at Stony Brook independently exhibited macroscopic quantum superpositions at the same time.) Over the past several years, Mooij and other researchers have built and exhibited the coherent control of superconducting qubits. Simple quantum computers consisting of several coupled superconducting qubits are currently being built and tested. I am now working in Japan with Jaio-Shen Tsai, Yasunobu Nakamura, and Tsuyoshi Yamamoto at NEC, trying to perform the first simple quantum computations on superconducting qubits.

Over the last decade, I have been lucky enough to work with some of the world's greatest experimental scientists to build quantum computers and quantum communication systems. The degree of their intimacy with nature is not something I can truly comprehend, let alone hope to emulate. These experimentalists possess the deepest possible theoretical understanding of quantum mechanics—the understanding needed to forge brand-new ways of talking to atoms and photons and convincing them to do what they have never done before.

CHAPTER 7

The Universal Computer

Simulating the Universe

We've shown how the laws of physics can be used to perform quantum computations in an efficient fashion. Now, let's consider how a quantum computer can efficiently simulate the operation of the laws of physics.

"Quantum simulation" is a process in which a quantum computer simulates another quantum system. Because of the various types of quantum weirdness, classical computers can simulate quantum systems only in a clunky, inefficient way. But because a quantum computer is itself a quantum system, capable of exhibiting the full repertoire of quantum weirdness, it can efficiently simulate other quantum systems. Every part of the quantum system to be simulated is mapped onto a collection of qubits in the quantum computer, and interactions between those parts become a sequence of quantum logic operations. The resulting simulation can be so accurate that the behavior of the computer will be indistinguishable from the behavior of the simulated system itself.

Recall that if two information-processing systems can simulate each other efficiently, they are logically equivalent. Because the universe can perform quantum computation and a quantum computer can simulate the universe, the universe and a quantum computer have the same information-processing power: they are essentially identical.

Quantum simulation is one of the most remarkable experimental demonstrations of the power of quantum computation to date—and the application of quantum computation that is most relevant to understanding the computational universe. It is because they tend to be doing many things at once that quantum systems are hard to simulate classically. Simulating a single nuclear spin, which can do two things in quantum parallel, is not so bad, but 10 spins can be doing 1,024 things at once and 20 spins can be doing 1,048,576 things at once, etc.

In general, to follow the dynamics of a quantum system, a classical computer has to assign a subcomputation to each piece of the quantum wave function, but the number of things the quantum system is doing grows very rapidly as the size of the quantum system increases. To simulate the dynamics of even a relatively small quantum system consisting of 300 nuclear spins is, as we've established, entirely unmanageable.

But a quantum computer has no trouble performing many subcomputations in quantum parallel. In 1982, the Nobel laureate Richard Feynman proposed a hypothetical device he called a universal quantum simulator. To simulate 300 nuclear spins, the universal quantum simulator would require only 300 quantum bits. As long as you could program the interactions between the 300 qubits so that they imitated the interactions between the 300 spins, then the dynamics of the qubits could simulate the dynamics of the spins.

Feynman merely noted the possible existence of universal quantum simulators; he gave no clue as to how they might be built. In 1996, I showed that conventional quantum computers were themselves universal quantum simulators; that is, any desired set of quantum-mechanical interactions could be programmed into a quantum computer and the quantum simulation could then take place by the repeated application of quantum logic operations on the computer's qubits.* (Techniques for performing quantum simulation were also derived independently around the same time by Christof Zalka of the University of Bern and Stephen Wiesner of Tel Aviv University.) Moreover, I was able to show

---

*"Universal Quantum Simulators," *Science* 273, no. 5278 (Aug. 23, 1996): 1073–78.

that the quantum simulation was efficient, in that (a) the number of qubits needed to perform the simulation was equal to the number of bits in the system to be simulated, and (b) the number of ops the quantum computer required to perform the simulation was proportional to the various lengths of time over which the system was to be simulated.

Feynman conjectured, and I proved, that quantum computers could function as universal quantum simulators whose dynamics could be the analog of any desired physical dynamics. The quantum simulation takes place in a simple and direct way. First, map the pieces of the quantum system to be simulated onto sets of quantum bits; each part of the system to be simulated gets just enough qubits to capture its dynamics. Second, map the interactions between the parts of the system onto quantum logic operations that act on the qubits corresponding to the parts. The universal nature of quantum logic operations ensures that any desired dynamics can be captured by this mapping.

Quantum simulation is not just a theoretical notion; it has been performed experimentally, as in Peter Shor's factoring algorithm. Unlike Shor's algorithm, however, which has so far only been shown to factor 15, quantum simulation has been performed on a scale that could never be replicated by a classical computer. Over the past several years, David Cory's group at MIT has performed quantum simulations involving billions and billions of qubits. These quantum simulators are crystals of calcium fluoride (I like to call them "weapons-grade toothpaste") about a centimeter across, with an eerie light-purple color stemming from trace amounts of other types of atoms. Each crystal contains more than a billion billion atoms. Using the techniques of NMR quantum computation to manipulate the nuclear spins in the crystals, Cory has made these nuclear spins adopt a wide variety of interactions. Most represent interactions not found in nature. To simulate such artificial quantum dynamics on a conventional classical computer would require the computer to perform 2 raised to the billion billion power subcomputations—rather a tall order. Cory's quantum simulators are far more powerful than any classical computer is or ever could be.

Cory's quantum simulations are far and away the most impressive quantum computations to date. But when I first presented his results in my lectures, I was surprised to find that many in the audience objected to describing these massive quantum simulations as computations. The typical response was "That's not a computation; it's an experiment!" I had a hard time understanding this response. Of course Cory was doing an experiment, I agreed—an experiment in quantum information processing. This seemed to sway many in the audience. But even when they had agreed that Cory was performing a computation, they would accept only that it was an analog quantum computation. It was hard for them to regard these analog quantum computations as "digital" quantum computation, like the factoring and search algorithms.

How do analog and digital computers differ? *A classical analog computer manipulates continuous variables,* such as voltage. This is because classical variables, such as position, velocity, pressure, and volume, are continuous, so to simulate classical dynamics an analog computer has to be continuous, too. *A classical digital computer deals with discrete quantities,* because bits are discrete; a classical digital computer can deal with continuous quantities but only by discretizing them.

In a quantum computer, however, there is no distinction between analog and digital computation. Quanta are by definition discrete, and their states can be mapped directly onto the states of qubits without approximation. But qubits are also continuous, because of their wave nature; their states can be continuous superpositions. Analog quantum computers and digital quantum computers are both made up of qubits, and analog quantum computations and digital quantum computations both progress by arranging logic operations between those qubits. Our classical intuition tells us that analog computation is intrinsically continuous and digital computation is intrinsically discrete. As with many other classical intuitions, this one is incorrect when applied to quantum computation. Analog quantum computers and digital quantum computers are one and the same device.

Simulation vs. Reality

The question of the difference between simulation and reality is an ancient one. In the sixth century B.C., in the first lines of the Tao Te Ching, Lao-tzu wrote of the problem inherent in describing reality: "The way that can be followed is not the true way. The name that can be named is not the true name." The original Chinese text of the Tao Te Ching is highly compact and susceptible to 10,000 interpretations, but Lao-tzu seems to be suggesting that by naming things—by assigning meaning to words—we introduce artificial distinctions that fail to capture the underlying wholeness of the universe. (To reduce the thought to bumper-sticker level: "Don't say it. Be it.") Here is a less literal translation of the same passage, by the philosopher Archie Bahm: "Nature can never be completely described, for such a description of nature would have to duplicate nature." That is, a perfect description of the universe is indistinguishable from the universe itself.

Let's look at what happens when we apply Lao-tzu's dictum to a quantum computer that is simulating the universe. As we will see, the universe—or at least the accessible part of the universe—is finite in space and in time. All the pieces of the accessible part of the universe can in principle be mapped onto a finite number of qubits. Similarly, the physical dynamics of the universe, consisting of the interactions between those pieces, can be mapped onto logic operations that act on those qubits.

This is not to say that we know exactly how to do this mapping. We know how to map the behavior of elementary particles onto qubits and logic operations. That is, we know how the Standard Model of particle physics—a model describing our world to superb precision—can be mapped into a quantum computer. But we don't yet know how the behavior of gravity can be mapped into a quantum computer, for the simple reason that physicists have not yet arrived at a complete theory of quantum gravity. We do not know how to simulate the universe yet, but we may know soon.

Now apply the Tao Te Ching. A quantum computer that simulated the universe would have exactly as many qubits as there are in the universe, and the logic operations on those qubits would exactly simulate the dynamics of the universe. Such a quantum computer would be a physical embodiment of the Marquis Pierre-Simon de Laplace's demon: it would simulate the behavior of the universe as a whole. Such a quantum computation would constitute a complete description of nature, and so would be indistinguishable from nature. Thus, at bottom, *the universe can be thought of as performing a quantum computation.* Likewise, because the behavior of elementary particles can be mapped directly onto the behavior of qubits interacting via logic operations, *a simulation of the universe on a quantum computer is indistinguishable from the universe itself.*

The conventional view is that the universe is nothing but elementary particles. That is true, but it is equally true that the universe is nothing but bits—or rather, nothing but qubits. Mindful that if it walks like a duck and quacks like a duck then it's a duck, from this point on we'll adopt the position that since the universe registers and processes information like a quantum computer and is observationally indistinguishable from a quantum computer, then it *is* a quantum computer.

The History of the Computational Universe

I have not been able to find any descriptions of the universe as a computer before the twentieth century. The ancient Greek atomists certainly thought of the universe as being constructed out of the interaction of tiny pieces, but they did not explicitly think of their atoms as processing information. Laplace conceived of his demon, able to calculate the entire future of the universe, as an abstract being, not as the universe itself. (Laplace did not call his being a demon; I rather think he thought his being was divine.) Charles Babbage seems to have been unconcerned about using his calculating machine as a model for physical dynamics, as was Alan Turing, although Turing was interested in the origins of patterns and complexity and did significant research on the subject.

The first explicit description I found of the universe as a computer is in Isaac Asimov's great 1956 science fiction story, "The Last Question." In this story, human beings create a sequence of ever larger analog computers to help them explore first their galaxy and then others. (In one Web parody of the story, the computer, which Asimov called Multivac, is renamed Google.)

Early in the story, in the year 2061, Lupov and Adell, two of Multivac's human co-workers, have an argument about the future of the universe and decide to ask the computer whether humanity will survive tens of billions of years hence, when the stars have all burned out. Lupov says:

> "It all had a beginning in the original cosmic explosion, whatever that was, and it'll all have an end when all the stars run down. . . . [J]ust give us a trillion years and everything will be dark. Entropy has to increase to maximum, that's all." . . .
>
> It was Adell's turn to be contrary. "Maybe we can build things up again someday," he said.
>
> "Never."
>
> "Why not? Someday."
>
> "Never."
>
> "Ask Multivac."
>
> "You ask Multivac. I dare you. Five dollars says it can't be done."
>
> Adell was just drunk enough to try, just sober enough to be able to phrase the necessary symbols and operations into a question which, in words, might have corresponded to this: Will mankind one day without the net expenditure of energy be able to restore the sun to its full youthfulness even after it had died of old age? Or maybe it could be put more simply like this: How can the net amount of entropy of the universe be massively decreased?

Multivac fell dead and silent. The slow flashing of lights ceased, the distant sounds of clicking relays ended. Then, just as the frightened technicians felt they could hold their breath no longer, there was a sudden springing to life of the teletype attached to that portion of Multivac. Five words were printed: INSUFFICIENT DATA FOR MEANINGFUL ANSWER.

In the story, time moves on. As humans explore the galaxy, then other galaxies, then become immortal (it's science fiction, after all), successive versions of Multivac become ever more powerful, eventually permeating the entire fabric of the universe. Humans continue to ask the computer variants of the same question about how to reverse the second law of thermodynamics, and all are answered the same way. Finally, when all of human intelligence, together with everything else, has been subsumed into Multivac's final incarnation, the universal AC, the computer figures it out and says . . . "LET THERE BE LIGHT!"

Note that in Asimov's story, the universe gradually *becomes* a computer, rather than starting out as one. We are interested in how the universe *began computing from the beginning.* The connections between computation and physics began to be worked out in the early 1960s by Rolf Landauer at IBM. The idea that computation could take place in a way that respects the underlying information-preserving character of physical law was developed in the 1970s by Charles Bennett at IBM and Edward Fredkin, Tommaso Toffoli, and Norman Margolus at MIT. The idea that the universe might be a kind of computer was proposed in the 1960s by Fredkin and independently by Konrad Zuse, the first person to build a modern electronic computer. Fredkin and Zuse suggested that the universe might be a type of classical computer called a cellular automaton, consisting of a regular array of bits interacting with their neighbors. More recently, Stephen Wolfram has extended and elaborated Fredkin's and Zuse's ideas.

The idea of using cellular automata as a basis for a theory of the universe is an appealing one. The problem with this argument is that classical computers are bad at reproducing quantum features, such as entanglement. Moreover, as has been noted, it would take a classical computer the size of the whole universe just to simulate a very tiny quantum-mechanical piece of it. It is thus hard to see how the universe could be a classical computer such as a cellular automaton. If it is, then the vast majority of its computational apparatus is inaccessible to observation.

Physical Limits to Computation

Once you know about quantum mechanics and quantum computation, it is surprisingly simple to determine just how much computation any physical system can perform. Start from the fact that all physical systems register information. Consider an electron that can be detected in one of two places, "here" and "there." An electron that can be either here or there registers one bit of information. (As Rolf Landauer said, "Information is physical.") When the electron moves from here to there, its bit flips. In other words, whenever a physical system changes its state—whenever anything at all happens—the information that the system registered is transformed and processed. (Information processing is also physical.)

Where electrons can be, and how they move from here to there, is governed by the laws of physics. The laws of physics determine how much information a particular physical system can register and how fast that information can be processed. Physics sets the ultimate limits to how powerful computers can be.

In a paper entitled "Ultimate Physical Limits to Computation," I showed how the computational capacity of any physical system can be calculated as a function of the amount of energy available to the system, together with the system's size.* As an example, I applied these limits to

*Nature 406 (Aug. 31, 2000): 1047–54.

calculate the maximum computational capacity of a kilogram of matter confined to a liter of volume. A conventional laptop computer weighs about a kilogram and takes up about a liter of space. I call such a one-kilogram, one-liter computer "the ultimate laptop." Next time you think about buying a new laptop, you should first compare it to the ultimate one.

Figure 13. The Ultimate Laptop

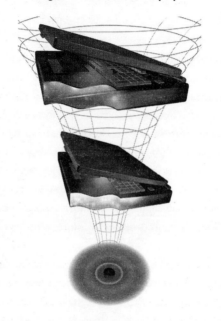

"The ultimate laptop" is a computer with a mass of one kilogram and a volume of one liter (the size of a conventional laptop computer) in which every elementary particle is put to use for the purposes of computation. The ultimate laptop can perform ten million billion billion billion billion billion logical operations per second on ten thousand billion billion billion bits.

Just how powerful is the ultimate laptop? The first fundamental limitation to computational performance comes from energy. Energy limits speed. For example, consider our one-bit electron moving from here to there. The more energy the electron has, the faster it can move from here to there and the faster it can flip its bit.

The maximum rate at which a bit can flip is governed by a useful theorem called the Margolus-Levitin theorem. Norm Margolus, as noted, is one of the pioneers in the physics of computation; working with his MIT advisor, Tommaso Toffoli, he showed that simple physical systems such as colliding atoms can perform universal digital computation. Lev Levitin of Boston University was one of the first scientists to use the laws of physics to calculate the capacity of communication channels such as fiber-optic cables for transferring information. They combined forces to publish their theorem in 1998.*

The Margolus-Levitin theorem says that the maximum rate at which a physical system (an electron, for example) can move from one state to another is proportional to the system's energy; the more energy available, the smaller the amount of time required for the electron to go from here to there. The theorem is very general. It doesn't care what system is registering and processing the information; it cares only how much energy the system has available to process that information. For example, consider the atoms and electrons in my computer. They are all at a temperature slightly higher than room temperature. Each atom and electron is jiggling around, and the amount of energy associated with a typical jiggle is the same for an atom as for an electron. The energy per jiggle is simply proportional to temperature, independent of whether it's an atom or an electron doing the jiggling. Thus the rate at which an electron inside a computer can move from one state to another—from here to there, or from 0 to 1—is the same as the rate at which an atom can move from one state to another. Electrons and atoms flip their bits at the same rate.

The Margolus-Levitin theorem supplies a method for calculating the maximum rate at which a bit can flip. Take the amount of energy available to flip the bit, multiply by 4, and divide by Planck's constant. The result is the number of times per second that the bit can flip. Applying

*Norman Margolus and Lev B. Levitin, "The Maximum Speed of Dynamical Evolution," *Physica D* 120 (1998): 188–95.

this method to the atoms and electrons in my computer, we find that each jiggling atom and electron flips its bit about 30 trillion (30×10^{12}) times per second.

The rate at which atoms and electrons flip their bits is typically much faster than the rate at which a conventional computer flips its bits. The computer upon which I type devotes a billion times more energy to charging and discharging the capacitors that register its bits than do atoms and electrons to jiggling and flipping theirs. But my computer operates 10,000 times more slowly than atoms do. The slowness of my computer doesn't contradict the Margolus-Levitin theorem. The theorem gives only an upper limit to how fast a bit can flip. A bit can flip more slowly than the maximum rate allowed by the theorem. A quantum computer, however, always flips its bits at the maximum rate.

The Margolus-Levitin theorem sets limits on the number of elementary operations (ops) that a bit can perform per second. Suppose we keep the amount of energy available for flipping bits the same, but now we divide the energy between two bits. Each of these two bits has half the energy of our original bit and can flip at half the rate. The total number of flips per second, however, remains the same.

If we divide the amount of available energy among ten bits, each will take ten times as long to flip, but the total number of flips per second will remain constant. Just as it is indifferent to the size of the system involved, the theorem doesn't care how the available energy is allocated. The maximum number of ops per second is given by the energy $E \times 4 \div$ Planck's constant.

The Margolus-Levitin theorem makes calculating the capacity of the ultimate laptop a breeze. The energy that the ultimate laptop has available for computation can be calculated using Einstein's famous formula $E = mc^2$, where E is the energy, m is the mass of the laptop, and c is the speed of light. Plugging in our ultimate computer's mass of one kilogram and the speed of light (300 million meters per second), we find that the ultimate laptop has 100 million billion (10^{17}) joules available with which to perform computations. Measured in a more familiar form of energy, the laptop has about 20 million million (2×10^{13}) kilocalories of available energy, equivalent to 100 billion candy bars. That's a lot of energy.

Another familiar measure of energy is the amount of energy released in a nuclear explosion. The ultimate laptop has twenty megatons (equivalent to 20 million tons of TNT) of energy available for computation, comparable to the amount of energy released by a large hydrogen bomb. In fact, when it is computing full throttle, using every available calorie to flip bits, the interior of the ultimate computer looks a lot like a nuclear explosion. The elementary particles that register and process information in the ultimate laptop are jiggling around at a temperature of a billion degrees. The ultimate laptop looks like a small piece of the Big Bang. (Packaging technology will have to make considerable progress before anyone will want to put the ultimate laptop on his lap.) Thus, the number of ops per second that our small but powerful computer can perform is a huge number, a million billion billion billion billion billion (10^{51}) ops per second. Intel has a long way to go.

Just how far does Intel have to go? Recall Moore's law: over the past half century, the amount of information that computers can process and the rate at which they process it has doubled every eighteen months or so. A variety of technologies—most recently, integrated circuits—have enabled this exponential increase in information-processing power. There is no particular reason that Moore's law should continue to hold year after year; it is a law of human ingenuity, not of nature. At some point, Moore's law will break down. In particular, no laptop can calculate faster than the ultimate laptop described above.

At its current rate of progress, how long will it take the computer industry to produce an ultimate laptop? The power of computers doubles every year and a half or so. Over fifteen years, it doubles ten times, going up by a factor of 1,000. Thus, current computers are a billion times faster than the giant, lumbering electromechanical computers of fifty years ago. Working flat out, current computers perform on the order of a trillion logical operations per second (10^{12}). So (if Moore's law can be sustained until then) we should be able to buy an ultimate laptop in the store by 2205.

The amount of energy available for computation limits the speed of the computation. But speed of computation isn't the only specification you're concerned with when you're buying a new laptop. Equally important is the amount of memory space. What is the capacity of the ultimate hard drive?

The interior of the ultimate laptop is jammed with elementary particles jiggling around like mad at a billion degrees. The same techniques that cosmologists use to measure the amount of information present during the Big Bang can be used to measure the number of bits registered by the ultimate laptop. The jiggling particles of the ultimate laptop register about 10,000 billion billion billion bits (10^{31}). That's a lot of bits—far more than the amount of information stored on the hard drives of all the computers in the world.

How long will it take the computer industry to reach the memory specifications of the ultimate laptop? In fact, Moore's law for memory capacity is currently moving faster than Moore's law for speed of computation. The capacity of hard drives doubles in a little over a year. At this rate, it will take only about seventy-five years to produce the ultimate hard drive.

Of course, Moore's law can be sustained only so long as human ingenuity can continue to overcome obstacles to miniaturization. It's hard to make wires, transistors, and capacitors smaller, and the smaller you make the components of computers, the harder they are to control. Moore's law has been declared dead many times in the past, each time because of some such knotty technical problem that seemed unsolvable. But each time it has been declared dead, clever engineers and scientists have found ways to cut the technical knot. Moreover, as we've discussed, we have hard experimental evidence that the components of computers can be miniaturized to the size of atoms. Existing quantum computers already store and process information at the atomic scale. At the current rate of miniaturization, Moore's law is not set to reach the atomic scale for another forty years, so there's hope for the old law yet.

The Computational Capacity of the Universe

Now that we know how much computation can be performed by a piece of matter that can sit on our lap, let's turn to more powerful computers—like the one envisioned by Isaac Asimov in "The Last Question," a computer the size of the cosmos itself. Suppose all the matter and energy in the cosmos were put into service to perform computations. How powerful would the resulting computer be? The power of the cosmological computer, consisting of everything in the universe, can be calculated using the same techniques with which we examined the power of the ultimate laptop.

First of all, energy limits speed. The amount of energy in the universe has been determined to a fairly high degree of accuracy. Much of it is locked up in the mass of atoms. If we count the atoms in all the stars and all the galaxies, adding the matter in interstellar clouds, we find that the overall average density of the universe is about one hydrogen atom per cubic meter.

There are other forms of energy in the universe, as well. For example, light contains energy (though far less than the energy contained in atoms). The rotation rates of distant galaxies suggest the existence of further, invisible sources of energy. The forms they take are unknown; candidates include such whimsically named constructions as wimps, winos, and machos. The anomalous acceleration of the universe's expansion suggests the presence of yet another form of energy, currently dubbed "quintessence." The total amount of energy in these exotic forms appears to be no more than ten times greater than the energy in directly observable matter, which doesn't make much difference in the total amount of computation the universe can perform.

Before we calculate the computational power of the universe, let's be clear about what it is we are calculating. Current observational evidence suggests that the universe is spatially infinite, extending forever in all directions. In a spatially infinite universe, the amount of energy in the universe is also infinite; consequently, the number of ops and number of bits in the universe is infinite, too.

But observation also tells us that the universe has a finite age: it is slightly less than 14 billion years old. Information can't travel any faster than the speed of light. Because the universe has a finite age and because the speed of light is finite, the part of the universe about which we can have information is also finite. The part of the universe about which we can have information is said to be "within the horizon." Beyond the horizon we can only guess as to what is happening. The numbers we will calculate represent the amount of computation that can take place within the horizon. Information processing occurring beyond the horizon cannot affect the result of any computation performed in the observable universe since the Big Bang. So when we calculate "the computational capacity of the universe," what we are really calculating is "the computational capacity of the universe within the horizon."

As time passes, the horizon expands, at three times the speed of light. When we look through a telescope, we also look backward in time, and the most remote objects we can see appear as they were a little under 14 billion years ago. In the intervening time, because of the expansion of the universe, those objects have moved even farther away, and right now they are 42 billion light-years away from us. As the horizon expands, more and more objects swim into view, and the amount of energy available for computation within the horizon increases. *The amount of computation that can have been performed within the horizon since the beginning of the universe increases over time.*

The horizon is 42 billion light-years away. On average, every cubic meter of the universe within the horizon contains a mass of about one hydrogen atom. Each hydrogen atom contributes energy $E = mc^2$. Toting up all the energy in the universe, we find that the universe contains about 100 million billion billion billion billion billion billion billion (10^{71}) joules of energy. Most of this is free energy, available for doing work or performing computation. That's a lot of calories! To eat that much, you'd have to be the size of the universe itself.

To get the maximum rate at which the universe can process information, then, apply the Margolus-Levitin theorem: take the amount of energy within the horizon, multiply by 4, and divide by Planck's constant. The result is that every second, a computer made up of all the energy in the universe could perform 100,000 googol (10^{105}) operations. Over the 14 billion years the universe has been around, this cosmological computer could have performed about 10,000 billion billion googol (10^{122}) ops.

By comparison, let's look at the number of ops that have been performed by all the computers on Earth since computers were first constructed. Because of Moore's law, half this computation has taken place in the last year and a half. (Whenever you have a process that doubles in capacity every year and a half, half that capacity has been generated in the last year and a half.) There are somewhat fewer than a billion computers on Earth. The clock rate of these computers is about a billion cycles per second (i.e., a gigahertz), on average. During each clock cycle, a typical computer performs somewhat fewer than 1,000 elementary operations. A year consists of about 32 million seconds. Over the last year and a half, then, all the computers on Earth have performed somewhat fewer than 10 billion billion billion (10^{28}) ops. Over the entire history of computation on Earth, computers have performed no more than twice this number of ops.

How many bits of memory space are available to the cosmological computer? Once again, the amount of memory space available is determined by counting the number of bits registered by every atom and every photon. Just as in the calculation of the memory space of the ultimate laptop, this number of bits can be counted using techniques developed by Max Planck a hundred years ago. The result is that the cosmological computer could store 100 billion billion billion billion billion billion billion billion billion billion (10^{92}) bits of information—far greater than the information registered by all of the computers on Earth.

The somewhat fewer than a billion earthly computers each have somewhat fewer than 1,000 billion (10^{12}) bits of memory space, on average, so taken together, they register fewer than 1,000 billion billion (10^{21}) bits.

The cosmological computer can have performed 10^{122} ops on 10^{92} bits. Those are big numbers, but I can think of bigger. In fact, when I first calculated the number of ops a computer the size of the universe could have performed, my initial reaction was "Is that all?"

Yes, that's all. No computer can have computed more than that, in the whole history of the universe. But it's also enough. Because of the power of quantum computers to simulate physical systems, a quantum computer that can perform 10^{122} ops on 10^{92} bits has enough power to compute everything we can observe. (If you take into account not only the bits that can be stored on elementary particles but those that can be stored using quantum gravity, soon to be described, there might be more bits—10^{122} bits, to be exact.) These numbers of ops and bits can be interpreted in three ways:

1. They give upper bounds to the amount of computation that can have been performed by all the matter in the universe since the universe began. As noted, the laws of physics impose fundamental limits on the speed of computation and the number of bits available. The speed of computation is limited by the amount of energy available, and the number of bits is limited by this energy together with the size of the system doing the computation. The size of the universe and the amount of energy in it are known to a fairly high degree of accuracy. No computer that obeys the laws of physics could have computed more.

2. They give lower bounds to the number of ops and bits required to simulate the universe with a quantum computer. Earlier, we saw that quantum computers are particularly efficient at simulating other quantum systems. To perform such a simulation, a quantum computer needs at least the same number of bits as the system to be simulated. In addition, to simulate each elementary event that occurs in the simulated system— for example, each time an electron moves from here to there—the quantum computer requires at least one op. A quantum computer that

simulates the universe as a whole must have at least as many bits as there are in the universe and must perform at least as many ops as the number of elementary events (or ops) that have occurred since the universe began.

3. The third interpretation is more controversial. If you choose to regard the universe as performing a computation, it could have performed 10^{122} ops on 10^{92} bits since its beginning. Whether or not you choose to regard the universe as performing a computation is to some degree a question of taste. To say the universe has performed 10^{122} ops requires you to define an op in terms of fundamental physical processes. In a computer, an op occurs when the computer flips a bit. (In some logic operations, such as an AND operation, the computer flips a bit or not, depending on the state of several other bits.) Here, we'll say that a physical system performs an op whenever it applies enough energy for enough time to flip a bit. With this simple physical definition of an op, the number of ops performed by any physical system, including the universe, can be calculated using the Margolus-Levitin theorem.

As time goes on, our horizon expands and the amount of energy available to register bits of information and perform computation increases. The total number of ops performed and the number of bits grow as a function of the age of the universe. In the Standard Model of cosmology, the total amount of energy within the horizon grows in direct proportion to the age of the universe. Since the rate of information processing is proportional to the energy available, the number of ops per second that the universe can perform within the horizon is also growing in proportion to the age of the universe. The total number of ops the universe has performed since the beginning is proportional to the number of ops per second times the age of the universe; that is, the total number of ops the universe has performed in the entire time since the Big Bang is proportional to the square of that time.

Similarly, conventional cosmology dictates that the number of bits within the horizon grows as the age of the universe raised to the three-quarters power. The information-processing power of the universe grows steadily with time. The future looks rosy.

So What?

We know how the universe is computing. We know how much the universe is computing. "So what?" you may ask. "Just what does this picture of the universe as a quantum computer buy me that I didn't already have?" After all, we have a perfectly good quantum-mechanical theory of elementary particles. So what if these particles are also processing information and performing computations? Do we really need a whole new paradigm for thinking about how the universe operates?

These are reasonable questions. Let's start with the last. The conventional picture of the universe in terms of physics is based on the paradigm of the universe as a machine. Contemporary physics is based on the mechanistic paradigm, in which the world is analyzed in terms of its underlying mechanisms; in fact, the mechanistic paradigm is the basis for all of modern science. A beautiful expression of it can be found in the opening paragraphs of Thomas Hobbes's *Leviathan,* his massive treatise on the political state:

> Nature, the art whereby God hath made and governs the world, is by the *art* of man, as in many other things, so in this also imitated, that it can make an artificial animal. For seeing life is but a motion of limbs, the beginning whereof is in some principal part within; why may we not say, that all *automata* (engines that move themselves by springs and wheels as doth a watch) have an artificial life? For what is the *heart,* but a *spring;* and the *nerves,* but so many *strings,* and the *joints,* but so many *wheels,* giving motion to the whole body, such as was intended by the artificer? *Art* goes yet further, imitating that rational and most excellent work of nature, *man.*

Paradigms are highly useful. They allow us to think about the world in a new way, and thinking about the world as a machine has allowed virtually all advances in science, including physics, chemistry, and biology. The primary quantity of interest in the mechanistic paradigm is energy.

This book advocates a new paradigm, an extension of the powerful mechanistic paradigm: I suggest thinking about the world not simply as a machine, but as *a machine that processes information*. In this paradigm, there are two primary quantities, energy and information, standing on an equal footing and playing off each other.

Just as thinking about the body in terms of clockwork allowed insight into physiology (and in Hobbes's case, into the inner workings of the body politic), the computational universe paradigm allows new insights into the way the universe works. Perhaps the most important new insight afforded by thinking of the world in terms of information is the resolution of the problem of complexity. The conventional mechanistic paradigm gives no simple answer to the question of why the universe in general, and life on Earth in particular, is so complex. In the computational universe, by contrast, the innate information-processing power of the universe systematically gives rise to all possible types of order, simple and complex.

A second insight provided by the computational universe pertains to the question of how the universe began in the first place. As noted, one of the great outstanding problems of physics is the problem of quantum gravity. In the beginning of the twentieth century, Albert Einstein proposed a beautiful theory for gravity called general relativity. General relativity, one of the most elegant physical theories of all time, accounts for many of the observed features of the universe at large scales. Quantum mechanics accounts for virtually all observed features of the universe at small scales. But to give a full picture of how the universe began, back when it was new, tiny, and incredibly energetic, requires a theory that unifies general relativity and quantum mechanics, two monumentally useful and undeniably correct theories that are basically incompatible.

There have been many valiant attempts to construct a quantum-mechanical theory of gravity. A clear summary of these efforts can be found in the theoretical physicist Lee Smolin's 2001 book *Three Roads to Quantum Gravity*. But none of these roads has yet reached its destination. Quantum computation provides a "fourth road," if you will. As with the other approaches, lots of roadwork remains to be done. And at any point in the development of such a theory, a fatal collision with experiment or observation can turn the theory to roadkill. Nonetheless, here is a map to the quantum-computational road to quantum gravity.

Quantum Computation and Quantum Gravity

Once you have grasped how quantum computations work, it is only a short distance to understand how general relativity works and how quantum computation could give rise to a unified theory of gravity and elementary particles. To see how quantum computation leads to general relativity, think of a circuit diagram for a quantum computation.

Figure 14. A Quantum Circuit Model for Spacetime

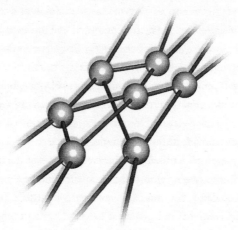

The fabric of spacetime in the computational universe is woven out of gates and wires. At each gate, two qubits interact: the gates are connected by wires that map out the paths that qubits take as they come together, interact, and move apart.

This circuit diagram describes what happens to the qubits in the quantum computation. The qubits go down "quantum wires" that take them to logic gates, where they interact. Additional wires then take them to other logic gates, where they interact with other qubits. Any quantum computation can be built up from these simple elements. The circuit diagram specifies the computation by giving its causal structure (the wires) together with its logical structure (the logic gates). The causal structure and the logical structure completely specify the quantum computation.

To construct a quantum theory of gravity from quantum computation, we need to show that quantum computations encompass the concept of space and time, together with the quantum matter that inhabits that space and time—and that Einstein's theory of general relativity can be derived from quantum computation. The derivation of gravity from quantum computation should specify how gravity reacts to quantum-mechanical matter and how the behavior of quantum-mechanical matter reacts to gravity. To be of any use, the theory should be predictive; that is, it should allow us to make retrodictions about what happened at the first instant of the universe and predictions about what happens when black holes evaporate—about the ultimate future of the universe.

This is a tall order, and we're certainly not going to solve all those problems here and now. The quantum-computational approach to the universe is an ongoing research program, not a solution to all the problems of physics (though we'll try to solve a few of them here).

General relativity is a theory of space and time and their interaction with matter. Each possible configuration of space and time interacting with matter is called a spacetime. Our universe is a particular spacetime.

In the computational-universe paradigm, the concepts of space and time, together with their interaction with matter, are to be derived from an underlying quantum computation. That is, each quantum computation corresponds to a possible spacetime—or more precisely, a quantum superposition of spacetimes—whose features are derived from the features of the computation. Our first goal is to show that the resulting spacetime obeys Einstein's theory of general relativity. Then we'll look at the predictions our theory makes for the computational universe.

Imagine the quantum computation as embedded in space and time. Each logic gate now sits at a point in space and time, and the wires represent physical paths along which quantum bits flow from one point to another. The first feature to note is that there are many ways to embed the quantum computation in space and time. Each quantum logic gate can be put down at any point where there is not another quantum logic gate, and the wires can squiggle all over the place to connect the logic gates. What happens to quantum information in the computation is independent of how the quantum computation is embedded in the spacetime. In the language of general relativity, the dynamical content of the quantum computation is "generally covariant"—that is, the quantum computation "doesn't care" how it is embedded in space and time as long as the qubits interact with one another in the right sequence.

The fact that a quantum computation doesn't care how it is embedded in spacetime means that the spacetime derived from the quantum computation obeys the laws of general relativity. Why? Because Einstein derived the laws of general relativity by requiring that those laws don't care how the underlying physical dynamics of matter is embedded in spacetime. Under the proper assumptions, general relativity is the only theory of gravity that is generally covariant.

The explicit verification that the spacetime derived from the quantum computation obeys the laws of general relativity is somewhat mathematical, but can be summarized as follows. The wiring diagram for the quantum computation dictates where information can go; it supplies a causal structure for spacetime. But general relativity tells us that the causal structure of spacetime fixes almost all features of the spacetime; just about the only feature that remains to be fixed are local length scales.

It's straightforward to see why local length scales are needed to determine the full structure of spacetime. Suppose I measure a distance here at MIT using a stick marked off into equal subunits. I measure the distance along MIT's "infinite corridor" (a very long but finite corridor that runs the length of the Main Building, in which my office sits). I find that it is twenty-five units long. Now I send you an e-mail message, wherever you are: "The infinite corridor is twenty-five units long." This e-mail message conveys no information to you about the infinite corridor's actual length unless you know the length of the unit I'm using. To convey to you the size of the unit, we need to establish a common standard of length. So if I tell you that my unit is equal to 1,650,763.73 wavelengths of orange-red light emitted by a krypton-86 atom (i.e., my unit is 10 meters), and if you have a krypton-86 atom, then you now know how long the infinite corridor is in terms of your local length scale. Nowadays, since time can be measured more accurately than length, the meter is defined as $\frac{1}{299,792,458}$ of the distance light travels in a second. If you prefer, I can define my unit to be 10 times $\frac{1}{299,792,458}$ of the distance light travels in a second (i.e., my unit is still 10 meters). Now you know how long the infinite corridor is—if you have some light and a clock capable of measuring small fractions of a second, that is.

Now return to the computational universe. Once the causal structure of the quantum computation has been specified, the only features of spacetime that remain to be fixed are local length scales, and these are to be fixed in terms of the wavelike properties of the local quantum-mechanical matter. The "matter" in the computational universe arises out of the quantum logic gates. Recall that any form of quantum-mechanical matter that arises out of local interactions can be simulated or constructed out of quantum logic gates. The quantum bits make up a sort of "quantum computronium," a computational form of matter capable of behaving like any elementary particle. Like a particle, each quantum logic gate corresponds to a wave, which wiggles up and down some number of times as the quantum bits are transformed by the quantum logic gate. The number of wiggles in the logic gate's wave is called the action of the logic gate.

As we follow the qubits through the computation, they accumulate action. The total action is just the total number of wiggles undergone by all the qubits in the course of the computation. It's a well-known fact of mechanics, both classical and quantum, that the behavior of any physical system is completely determined by its action. The action of the quantum logic gates completely determines what happens during the computation. As I like to put it, the action is where the action is.

Einstein's equations relate the geometry of spacetime to the behavior of the matter in it. That geometry tells matter where to go, and the matter tells the geometry how to curve. Einstein's equations relate the curvature of spacetime at a given point to the action at that point—in our case, to the number of wiggles in the wave of the quantum logic gate. We have to verify that Einstein's equations hold for our computational picture of gravity.

To fully fix the curvature requires us to choose local length scales. Once these scales are chosen, the structure of the computational spacetime is entirely fixed. It is straightforward to show that the local length scales can always be chosen so that the resulting spacetime obeys Einstein's equations for general relativity. This agreement with general relativity is no accident. (Because the quantum computation doesn't care how it is embedded in spacetime, our theory is automatically covariant. As a result, once it has been embedded in spacetime, the quantum computation has essentially no choice but to obey Einstein's equations.)

Einstein challenged John Wheeler to sum up general relativity in a simple phrase. Wheeler rose to the challenge: "Matter tells space how to curve," he said, "and space tells matter where to go." Let's rephrase Wheeler's dictum for the computational universe: "Information tells space how to curve; and space tells information where to go." In the computational universe, space is filled with "wires," paths along which information flows. The wires tell information where to go. The wires meet at quantum logic gates, where that information is transformed and processed. The quantum logic gates, in turn, tell space how much to curve at that point. The structure of spacetime is derived from the structure of the underlying computation.

The computational universe picture of quantum gravity predicts a variety of features of the universe around us. It gives a straightforward picture of how spacetime responds to the presence of quantum-mechanical matter. It can be used to calculate how quantum fluctuations in the early universe programmed the density of matter and the locations of future galaxies. It supports models for black-hole formation and evaporation. The interacting qubits of the underlying quantum computation are perfectly capable of reproducing the phenomenology of the Standard Model for elementary particles. In other words, quantum computation represents what physicists like to call a Theory of Everything (or TOE). Mindful that theories of everything are too often theories of almost nothing, I prefer to call it a *potential* theory of everything (or PTOE). The motto of this potential theory of everything is, to paraphrase John Wheeler, "It from qubit!"

The computational universe paradigm for the interaction of quantum mechanics with general relativity represents a distinct road to quantum gravity. It travels through a very different landscape than Smolin's three roads, but its final goal is the same. This paradigm is a work in progress. It makes explicit predictions for the behavior of the early universe and for processes such as black-hole evaporation. These explicit predictions can be tested by observation—for example, by detailed observations of the cosmic microwave background, the remnant radiation of the Big Bang. Time will tell whether the computational-universe paradigm is a road to understanding quantum gravity or whether collision with observation and experiment will kill it.

Despite the inevitable uncertainty of science in the making, the theory of general relativity as a consequence of quantum computation has passed a scientific milestone that has not yet been passed along any of the other three roads. Because quantum computation so easily encompasses and reproduces quantum dynamics, the computational-universe theory of quantum gravity combines general relativity and the Standard Model of elementary particles in a straightforward and self-consistent way. This accomplishment suggests that if we follow the road of the computational universe, it may well lead us to our goal, to understand the universe, and everything in it, in terms of how it processes information.

CHAPTER 8

Complexity Simplified

Making Things Complex

The primary consequence of the computational nature of the universe is that the universe naturally generates complex systems, such as life. Although the basic laws of physics are comparatively simple in form, they give rise, because they are computationally universal, to systems of enormous complexity. Besides encompassing the Standard Model of elementary particles and leading at least part of the way to a theory of quantum gravity, the computational universe provides an explanation for one of the most important features of the universe: its complexity. In the beginning, the universe was simple. Now it isn't. So what happened?

As far as astronomers and cosmologists can tell, the universe at the moment of the Big Bang was a rather uncomplicated affair. It was very hot everywhere, and everywhere looked the same; that is, the initial state of the universe was characterized by regularity, symmetry, and simplicity. Look to the skies now and the picture is very different. There are planets, stars, galaxies, clusters of galaxies, superclusters of galaxies. The universe is clumpy and highly asymmetric. Now look out your window: plants, animals, people, cars, buildings. Life on Earth is far from simple. Just how did the universe manage to become so complex? The results presented in this book allow us to give a scientifically testable answer to this question.

We have already established a framework for describing the operation of the universe in terms of quantum information processing. We've seen that a quantum computer can simulate the universe in an efficient way, and that the universe is thus observationally indistinguishable from a quantum computer. To buttress these theoretical arguments, we have strong empirical evidence that the universe supports computation: I am writing this on a computer that obeys the laws of physics. Therefore the laws of physics support digital computation, and if I can keep on buying new memory space, my computer is a universal digital computer. Clearly, whatever the underlying laws of physics are, they allow the construction of computers at relatively macroscopic scales.

There is also strong evidence that the universe supports computation at the most microscopic levels. The quantum computers my colleagues and I are building bear witness to the ability of matter and energy to perform computation at the smallest scales: we are able to control the behavior of atoms, electrons, and photons in a precise fashion. Regardless of what form matter and energy take at ever smaller scales, as long as they obey the laws of quantum mechanics they can be used to compute. In the cosmological universal computer (the universal computer consisting of the universe itself), every atom is a bit, every photon moves its bit from one part of the computation to another, and every time an electron or a nuclear particle changes its spin from clockwise to counterclockwise, its bit flips. Until we obtain a full quantum theory of all fundamental physical phenomena, including gravity, we will not be able to verify the computational mechanics of the universe in detail. But we can (and do) hope that one day such verification will be possible.

If the universe is indeed a quantum computer, this presents an immediate explanation of the complexity we see around us. To understand why the computational ability of the universe essentially guarantees its current complexity, let's return to the story of Ludwig Boltzmann and the typing monkeys. Recall that Boltzmann claimed that the complexity of the universe arose from chance. What we see around us, he asserted, is just the result of a statistical fluctuation, no different from the outcome of a long sequence of coin tosses. At first, this might seem to be an appealing explanation: in an infinitely long sequence of coin tosses, any desired finite sequence of heads and tails will eventually occur, including a representation of any desired text or mathematical expression encoded in binary form. Such a random construction of texts is the basis of Jorge Luis Borges's story "The Library of Babel," which describes a fictional library containing all possible texts. The story of monkeys typing *Hamlet* is a variant on this version of the origins of complexity.

But Boltzmann's explanation of the origins of complexity is demonstrably incorrect. The vast majority of sequences of coin flips exhibit no apparent order or complexity. If complexity arises only at random, then no matter how much ordered or complex behavior has been revealed so far, what occurs next will be random. No matter how far into *Hamlet* a

Figure 15. Monkeys Typing

a

Monkeys typing on typewriters (figure 15a) generate gibberish. When the same monkeys type into computers (figure 15b), the gibberish they generate is interpreted as a computer program by the computers, and generates complex structures.

monkey may get, its next keystroke is likely to be a mistake. In a universe where everything arises at random, our next breath is definitely our last, as our atoms immediately reconfigure to a random state. (In Borges's story, a book taken off the shelves at random will be gibberish, and the call number for any book is as long as the book itself. The Library of Babel is useless.)

Boltzmann realized that his explanation of the universe as statistical fluctuation was wrong and apparently did not pursue the question further. However, there is still a germ of truth in Boltzmann's idea. As in chapter 3, to obtain a more plausible explanation of the origins of complexity, we imagine that the monkeys are typing not on typewriters but

Figure 15. Monkeys Typing *(continued)*

b

into computers. The computers then interpret the gibberish as a set of instructions to execute, written in, say, Java. What comes out of the computers? Garbage in, garbage out: with high probability the monkey-wielded computers will produce an error message. But sometimes one of them will produce something more interesting. The probability of the monkeys generating any given program decreases rapidly with its length. But short programs exist for producing a wide variety of interesting outputs.

At the beginning of the 1960s, computer scientists developed a detailed theory of how likely it was for a randomly programmed computer to produce interesting outputs. That theory is based on the idea of "algorithmic information."

Algorithmic Information

Algorithmic information is a measure of how hard it is to represent a text or a bit string using a computer. The algorithmic information content of a text or a bit string is equal to the length, in bits, of the shortest computer program that produces that text or bit string as output.

In chapter 2, we saw that computer languages provide a method for assigning meaning to strings of bits. Using such a language, these strings can be interpreted as instructions that tell a computer to produce a particular output. However, for any desired output, there are many languages to choose from, and typically many computer programs able to produce the sought-after result. For example, there are many different programs that produce as their output the first million digits of π. Significantly, though, they are not all of the same length. One program simply says "PRINT 3.1415926 . . ." (where the " . . ." is a list of the next 999,992 digits). This program is straightforward, but long. A shorter, more sophisticated program would specify a particular method of *calculating* those million digits. For example, such a program could follow the example of the ancient Greeks and approximate the circumferences of smaller and smaller circles as a sequence of straight lines. A program to calculate the digits of π by this procedure might be no more than a few hundred instructions long.

For any desired output, the shortest of the available programs that produce it is particularly interesting. For any number, the "algorithmic information content" is defined as the length in bits of the shortest computer program enabling the computer to print out that number. This shortest computer program can be thought of as the most compact representation of the number in the given computer language.

Algorithmic information content was discovered independently at the beginning of the 1960s by the Cambridge, Massachusetts, scientist Ray Solomonoff, the Russian mathematician Andrei Nikolaevich Kolmogorov, and Gregory Chaitin, then an eighteen-year-old student at the City College of New York. These three researchers noted that algorithmic information content provided in some ways a more satisfying measure of information than the length of a number in bits (which is another way of describing the number's information content) because algorithmic information respects the intrinsic mathematical regularities of a number in a way that the length in bits fails to grasp.

For most numbers, the algorithmic information content is close to the length of the number in bits. It can't be much *longer* than the length of the number, since any number—e.g., 01110110101110111011101—can be produced by a program that says "PRINT 01110110101110111011101." And for most numbers, the algorithmic information content can't be much *shorter* than the number, either, simply because there are far fewer short programs than there are long numbers. We can ask, for example, how many twenty-bit numbers could be produced by ten-bit programs. There are 2^{20} (that is, 1,048,576) twenty-bit numbers but only $1,024 = 2^{10}$ possible ten-bit programs. So, at most, one out of 1,024 twenty-bit numbers could be produced by a ten-bit program.

The numbers that *can* be produced by short programs are those that have mathematical regularities. *Pi* is one such number, and so is the number consisting of a billion 1's, which can be produced by a program that says "PRINT '1' one billion times." But, as noted, most numbers don't have significant mathematical regularities. Most numbers are effectively random.

The shortest computer program that will produce a number is always defined relative to a given computer language—Java, C, Fortran, BASIC. But its length does not depend strongly on which language is used. Most languages can produce the first million digits of π using only a few hundred instructions. In fact, a program to produce a number written in Fortran can be translated into a program to produce the same number written in Java by supplying a translating program. As a result, the shortest Java program for producing the first million digits of π is no longer than the shortest Fortran program plus the length of the translating program. As the number to be produced gets longer and longer, the length of the translating program becomes, relatively, smaller and smaller, adding comparatively little length to the algorithmic information content.

This translatability of computer programs is a central feature of computation. A program written in Fortran can always be translated into another program written in Java. This translatability is an aspect of the universal nature of computation. Another familiar universal aspect of computation is that the same programs—Microsoft Word, for example—can run on computers of different architecture. The wiring diagrams of a Macintosh and a PC are very different, as is the way in which a particular instruction is represented and executed on either machine. But both computers can run Word. When Word is loaded into a Macintosh, the program is translated (or "compiled") into a set of instructions the Macintosh understands, and the same is true of a PC. Despite these underlying differences, someone writing a Word document on a Mac will tap the same keys to write the same sentence as someone writing the same document on a PC.*

Algorithmic information is an appealing concept relying on the uni-

---

*Running Word on a PC is not completely equivalent to running it on a Mac, however; Word may run more slowly on one than on the other. Certain versions of Word run famously slowly on the Macintosh. Translations are accurate, but they aren't always efficient.

versality and translatability of computer languages. It allows bit strings with mathematical regularity to be expressed in a compact form. The shortest program to produce a bit string can be thought of as a compressed representation of the bit string.

Many bit strings in the real world have mathematical regularities and so can be compressed. For example, in English, the letters occur with different frequency: E is most common, followed by T, A, I, O, N, S, H, R, D, L, U (the top row of the bins of movable type used by typesetters). A program that recodes English so that E corresponds to a shorter code and Q to a longer one can compress English texts by a factor of about two.

Algorithmic Probability

Ray Solomonoff originally defined algorithmic information while looking for a formal mathematical theory of Occam's razor. The medieval philosopher William of Occam was interested in finding the simplest explanation for observed phenomena. *Pluralitas non est ponenda sin necessitate,* he declared: "Plurality should not be posited without necessity." Occam urged us to accept simple explanations for phenomena over complex ones. Occam would have scoffed, for example, at people who have postulated the existence of Martians in order to explain the regular lines observable on the surface of Mars. Geological faults or optical illusion would just as well explain these lines without requiring the presence of Martians. To declare the existence of Martians in order to explain the lines on Mars is to "multiply beings beyond necessity," or, simply put, to make things unnecessarily complicated. Occam's razor cuts away complex explanations by declaring simple ones to be a priori more plausible.

Solomonoff used algorithmic information content to make Occam's razor mathematically precise. Suppose we are given a set of data, expressed as a string of bits. We are looking for mechanisms that might plausibly have produced this bit string. Phrased in terms of computation, we are looking for computer programs that produce this bit string as output. Among these computer programs, Solomonoff declared, the shortest program is intrinsically the most plausible guess as to which one generated the string.

Just how much more plausible? In the 1970s, Gregory Chaitin and his colleague Charles Bennett at IBM began to talk about algorithmic information in terms of monkeys. Suppose a monkey types random strings of bits as input into a computer. The computer interprets the strings as programs written in a suitable language—Java, say. What is the probability that the computer outputs the first million digits of π? It's the same as the probability that the random strings typed by a monkey reproduces a Java program for calculating the first million digits of π. The probability of a monkey typing out the very first bit of such a program correctly is, of course, 50-50, or one-half. The probability of getting the first two bits right is 75-25, or one-quarter. The probability of getting the first 1,000 bits right is $\frac{1}{2}$ multiplied by itself 1,000 times, or $\frac{1}{2}^{1000}$, a very small number. Obviously, the longer the program, the more unlikely it is that a monkey will type it out correctly.

The probability that the random program the monkey inputs into the computer will give the first million digits of π as output is called the "algorithmic probability" of π. Since long programs are so much less likely to be typed correctly than short programs, the algorithmic probability is greatest for the shortest programs. The shortest program that can output a particular number is the most plausible explanation for how that number was produced.

To look at this from another angle, numbers produced by short programs are more likely to appear as outputs of the monkey's computer than numbers that can be produced only by long programs. Many beautiful and intricate mathematical patterns—regular geometric shapes, fractal patterns, the laws of quantum mechanics, elementary particles, the laws of chemistry—can be produced by short computer programs. Believe it or not, a monkey has a good shot at producing everything we see.

Algorithmically probable things are just those things that exhibit large amounts of regularity, structure, and order. In other words, whereas the typewriter monkey universe is just gibberish, the computer monkey universe contains, along with plenty of gibberish, some interesting features. Large parts of the computer monkey universe consist of structures that can be generated from simple mathematical formulae and concise computer programs. If the monkeys type into computers instead of typewriters, they produce a universe with a mix of order and randomness, in which complex systems arise naturally from simple origins—that is, they produce a universe suspiciously like our own. Simple programs together with lots of information processing give rise to complex outputs. Can this be an explanation for the complexity of our universe?

What is required to make this explanation testable? For the computational explanation of complexity to work, two ingredients are necessary: (a) a computer, and (b) monkeys. The laws of quantum mechanics themselves supply our computer. But where are the monkeys? What physical mechanism is injecting information into our universe, programming it with a string of random bits? We need, again, look no further than the laws of quantum mechanics, which are constantly injecting new information into the universe in the form of *quantum fluctuations*. In the early universe, for example, galaxies formed around seeds—places where the density of matter was a tiny bit higher than elsewhere. The seeds of galaxy formation were provided by quantum fluctuations: the average density of matter was everywhere the same, but quantum mechanics added random fluctuations that allowed galaxies to coalesce.

Quantum fluctuations are ubiquitous, and they tend to insert themselves at the points where the universe is most sensitive. Take biology, for instance. You get your DNA from your mother and father, but your exact sequence of DNA is produced by a process of recombination during the formation of your father's sperm cells and your mother's egg cells. Just which of your mother's and father's genes you get depends sensitively on chemical and thermal fluctuations during the recombination process, and these chemical and thermal fluctuations can be traced back to quantum mechanics. Quantum accidents programmed your DNA to be different from that of your brother or sister. You and I, and the differences between us, came from quantum accidents. And so, from quantum seeds, came the universe itself. *Quantum fluctuations are the monkeys that program the universe.*

Randomness arises in the computational universe because the initial state of the universe is a superposition of different program states, each one of which sets the universe down a different computational path and some of which result in complex and interesting behavior. The quantum-computational universe follows all of these computational paths simultaneously, in quantum parallel, and these computational paths correspond to the decoherent histories described earlier. Since the computational histories decohere, we can talk about them at the dinner table; one or another of these histories has actually occurred. One of these decoherent histories corresponds to the universe we see around us.

What Is Complexity?

The computational universe spontaneously gives rise to every possible form of computable behavior; anything it can be programmed to do, it does. Some of this behavior is orderly, some is random; some is simple, some is complex. But what is complexity, anyway?

Halfway through my Ph.D. in physics at Rockefeller University, I was almost expelled. I had come to Rockefeller because of its reputation for supporting independent work. After passing the qualifying exams, I was pursuing research on the role of information in quantum-mechanical systems and how quantum information processing might be related to a variety of fundamental processes in physics, including quantum gravity. In other words, I was doing just what I do now, some twenty years later. I had no immediate supervisor. One day in 1986 two professors and Heinz Pagels, the executive director of the New York Academy of Sciences, walked into my office. "Lloyd," they said, "you must stop working on this crazy stuff and switch to a topic that we understand. If you do not, you must leave Rockefeller."

This announcement came as a complete surprise. I knew I was working outside the bounds of the normal topics in the department. Most of the other graduate students were working on string theory, a highly abstract theory that was shaving space into multiple unvisualizable dimensions in an attempt to reconcile quantum theory with general relativity and thus explain all known fundamental physics. I couldn't for the life of me see why what I was doing was crazier than string theory.

But many other people were working on string theory, and at that point only a handful of people around the world were working on quantum information. I would later meet these people and work with them, but at the time I didn't even know who they were. So one immediate result of this meeting was that I knuckled under and agreed to solve two conventional problems in quantum field theory over the following few months. The best result of being on the verge of expulsion was that I got to work with Heinz Pagels. Heinz was a striking figure, fond of double-breasted pin-striped suits and patent leather boots. With his bouffant white hairdo and these Mafia duds, he resembled a slender John Gotti. He was a wild man of physics and he was willing to supervise me.

After four months, I had done both the problems I had been assigned. After eight months, I had convinced Heinz that looking at black-hole evaporation in terms of quantum information processing might not be such a bad idea. After a year, he was taking me on tours of the East Village to meet his even more outré friends, most of them in the world of performance art. He also introduced me to his wife, Elaine, the author of *The Gnostic Gospels,* a book that completely changed my ideas about the social nature of religion. I might have been headed for the profession of taxi driver, like so many other unemployed physics Ph.D.s, but now, at least, I was having some fun along the way.

The real turning point of our intellectual relationship came on the day Heinz walked into my office and said, "OK, Seth, how are we going to measure complexity?"

"We can't," I said. "Things are complex exactly when they defy quantification."

"Bullshit," said Heinz. "Let's try."

Asking how to measure complexity is like asking how to measure physics. There are many measurable quantities among the laws of physics—energy, distance, temperature, pressure, electric charge—but "physics" itself is not a measurable quantity. Similarly, in identifying the laws of complexity, we should expect there to be a number of measurable quantities that make up a complex system. I spent some months reading up on various methods of defining complexity. The first concept I looked at was computational complexity. The complexity of a computation is equal to the number of elementary logical operations that must be performed in the course of the computation. (A related concept, spatial computational complexity, is equal to the number of bits used in the course of the computation.) Computational complexity is not a measure of complexity so much as a measure of effort, or of the resources required to perform a given task. There are plenty of computations that take a long time and use up lots of space but do not produce anything very complex. As will be seen, computational complexity is an important ingredient in a good definition of complexity—but it is not itself a good definition.

Algorithmic information had also been proposed as a measure of complexity—in fact, Chaitin had originally called it "algorithmic complexity." But bit strings with high algorithmic information content don't look complex, they look random; indeed, algorithmic information has also been called algorithmic randomness. Moreover, bit strings with high algorithmic information content are easy to create: just flip a coin 100 times. The resulting bit string will likely have close to the maximum possible algorithmic information content. Heinz and I felt that a complex thing should be intricate, structured, and hard to reproduce. Things with high algorithmic information content may require lots of bits to describe, but most bit strings with high algorithmic information content are unstructured and easy to produce.

As I looked further and further, I found more and more definitions of complexity, each a variation on the theme of required effort or amount of information. Several years later, I gave a talk on these various measures of complexity at a conference sponsored by the Santa Fe Institute, which had been founded in the mid-1980s by George Cowan, Murray Gell-Mann, and a group of senior scientists at nearby Los Alamos who were interested in examining the rules that underlie and give rise to complex systems. The talk was titled "Thirty-one Measures of Complexity," the "thirty-one" being a convenient allusion to the number of flavors of ice cream offered by Baskin-Robbins. Although I had not published a paper with this title, word of the talk spread throughout the Internet, and for many years my paper on thirty-one measures of complexity was my most frequently requested work, despite the fact that it didn't exist. (I finally published the list as a paper a couple of years ago,* just so that I would no longer have to respond to all the e-mails requesting this nonexistent paper.) As it happened, between the talk and the publication of the paper, the number of measures of complexity on the list grew from thirty-one to forty-two. (The length of the new list prompted the writer John Horgan to claim, in his book *The End of Science,* that the science of complex systems was bankrupt, since researchers couldn't even agree on what complexity was, let alone conduct any significant research about it.)

The paper divides measures of complexity into four categories: first, measures (like algorithmic information) of how hard it is to describe something; second, measures (like computational complexity) of how hard it is to do something; third, measures of the degree of organization in a system; fourth, non-quantitative ideas associated with complexity (like self-organization or complex adaptive systems). Of the forty-two, those I find most interesting are measures that combine how hard it is to describe something, how hard it is to do something, and the degree of organization into a single measure. It is on this set of measures I will concentrate here.

The laws of physics describe trade-offs and relationships between measurable quantities, and the laws of complexity do the same. A partic-

*S. Lloyd, "Measures of Complexity: A Nonexhaustive List," *IEEE Cont. Syst. Mag.* 21, no. 4 (2001): 7–8.

ularly useful trade-off is that between information and effort. Consider algorithmic information as a measure of information content, and computational complexity as a measure of effort required. Consider the amount of effort required to produce a particular bit string—the first million digits of π, for example. These digits can be produced with relatively little computational complexity (only a few million logical operations, taking under a second on a conventional computer) by the million-plus-digit program that says "PRINT 3.1415926 . . ." Although we don't know the exact algorithmic information of the first million digits of π (recall that algorithmic information is uncomputable), we can always find an upper bound to that algorithmic information by producing short programs that compute the first million digits of π. For instance, a program that computes those digits using a mathematical technique called a continued fraction representation could have a length of less than 1,000 digits, but this compact program takes a lot longer to produce the first million digits of π than the simple but bulky PRINT program. It would take billions of logical operations to produce those million digits.

In the early 1980s, Charles Bennett proposed a simple definition of complexity that relies on the trade-off between information and effort. Following Solomonoff, Bennett identified the most plausible explanation of a bit string or data set with the shortest program that produced it. (If there were several programs that were almost as short as the shortest, Bennett included those as plausible explanations, too.) Then Bennett looked at the computational complexity of those short programs. He called this quantity—the effort required to produce the bit string from its most plausible explanation—"logical depth."

Of all the measures of complexity Heinz and I studied, logical depth was the most appealing. Bit strings that are obviously simple, like the string consisting of a billion 1's, have short fast-running programs that can produce them (e.g., "PRINT 1 ONE BILLION TIMES") and are logically shallow. Random bit strings (e.g., 11010101100010 . . . 011, a bit string I got by flipping a coin and calling heads 1 and tails 0) are plausibly produced by long fast-running programs (e.g., "PRINT 11010101100010 . . . 011") and are also logically shallow. By contrast, bit strings corresponding to the first million digits of π take a long time to produce from their shortest known programs and are logically deep. Logically deep bit strings possess large amounts of structure—structure that takes a long time to compute from the shortest possible program.

Heinz and I were very taken by Bennett's ideas on complexity. Heinz's only complaint was that this scheme was not physical enough. "Logical depth" referred to bit strings, computer programs, and logical operations. Heinz wanted a measure of complexity that referred to physical systems—energy and entropy. So he and I concocted a physical analog to logical depth, which we called "thermodynamic depth," to emphasize the connection to Bennett's work. Instead of being a property of bit strings, thermodynamic depth is a property of physical systems. Instead of identifying the most plausible way that a bit string was produced with the shortest program that produced it, Heinz and I looked directly at the most plausible way that *a physical system* was produced. Finally, instead of using the computational complexity—the number of logical operations—that went into producing a given bit string, we looked at the amount of physical resources needed to produce the given physical system—an atom, say, or an elephant.

The particular physical resource Heinz and I considered was related to entropy. Recall that entropy is measured in bits. Entropy consists of random, unknown bits. The opposite of entropy is called "negentropy." Negentropy consists of known, structured bits. A system's negentropy is a measure of how far away that system is from its maximum possible entropy. A living, breathing human being has lots of negentropy, as opposed to, say, a gas of helium atoms at uniform temperature, which has *no* negentropy. You can think of entropy as consisting of random, junky bits and negentropy as consisting of ordered, useful bits. The thermodynamic depth of a physical system is equal to the number of useful bits that went into assembling the system.

Because it is the acknowledged offspring of logical depth, thermodynamic depth shares many of logical depth's good qualities. Simple, regular systems that are easily assembled, such as salt crystals, are typically thermodynamically shallow. Fully random systems, such as our gas of helium atoms, generated by a straightforward random process such as heating, are also thermodynamically shallow. But intricate, structured systems, such as living systems, required a huge investment of useful bits over billions of years to assemble and are thermodynamically deep.

When applied to bit strings (for example, those produced by a randomly programmed quantum computer), thermodynamic depth is even closer to logical depth. The most plausible way a bit string can be produced is from the shortest program. Thus the thermodynamic depth of the bit string is the amount of memory space used by the quantum computer in producing the string; that is, the thermodynamic depth is the spatial computational complexity of the shortest program.

In the computational universe—in which each physical system does indeed correspond to a string of quantum bits and its behavior is programmed by random quantum fluctuations—thermodynamic depth and logical depth are complementary, closely related quantities. To nail down the analogy between thermodynamic depth and logical depth, we need to find a physical analog to the elementary logical operation, or op. The previous chapter defined just such an analog: every time a quantum wave wiggles, an op is performed. To give a physical analog of the number of ops that went into constructing a bit string, just count the number of wiggles that went into constructing a physical system.

Recall from the previous chapter that this number of wiggles is proportional to what in physics is called the action of the physical system. Action is the number of wiggles multiplied by Planck's constant. Action divided by Planck's constant is a good physical analog for the number of ops, i.e., for computational complexity. To estimate how hard it was to construct a given physical system, just look at the action that went into putting it together. ("The action is where the action is.")

The results of the previous chapter now allow us to estimate the logical depth and thermodynamic depth of the universe as a whole and so to put an upper bound on the depth of everything that it contains. The total amount of computational effort that went into putting the universe together is 10^{122} ops (the logical depth) performed on 10^{92} bits (the thermodynamic depth).

Effective Complexity

Logical and thermodynamic depth are not the only measures that quantify some aspect of complexity. Depending on which feature of a complex system you wish to characterize, there are other measures that are equally or even more useful. One such measure is the quantity known as "effective complexity," a measure of the amount of regularity in a system; this definition of complexity was originally proposed by Murray Gell-Mann. Over the last decade, Gell-Mann and I have worked to make the notion of effective complexity mathematically precise.

Effective complexity is a simple and elegant measure of complexity. Every physical system has associated with it a quantity of information—the amount required to describe the physical state of the system to the accuracy allowed by quantum mechanics. The basic way to measure something's effective complexity is to divide that amount into two parts: information that describes the regular aspects of the thing and information that describes its random aspects. *The amount of information required to describe a system's regularities is its effective complexity.*

In an engineered system, such as an airplane, the effective complexity is essentially equal to the length of the system's blueprint: it is the amount of information required to put the system together. In an airplane, for example, the blueprint specifies the shape of the wings and the chemical content and manufacturing procedure for the alloy from which the wings are made. Wing shape and alloy composition are *regularities* of the design; the bits that specify these features have to take on specific values if the airplane is to fly. These bits figure in the airplane's effective complexity. But the blueprint does not specify the position of each and every atom in the wing. The bits that specify just where each atom is at one or another point in time are accidents; they do not contribute to the flight-worthiness of the airplane, nor are they an indicator of its complexity.

As the example of an airplane suggests, complexity is a key issue in engineering. How can we engineer complex systems that are still robust in their behavior? The maxim we teach engineering undergraduates at MIT goes by the acronym KISS: Keep It Simple, Stupid! But what if the system you're engineering is itself complex, like an airplane? MIT has an entire division, the Engineering Systems Division, that brings together researchers from engineering, the hard sciences, and the social sciences to identify and solve problems of complex engineered systems. One promising technique for engineering complex systems is known as axiomatic design, an approach conceived by Nam Suh, the former head of MIT's Department of Mechanical Engineering. The idea of axiomatic design is to minimize the information content of the engineered system while maintaining its ability to carry out its functional requirements. Properly applied, axiomatic design results in airplanes, software, and toasters all just complex enough, and no more, to attain their design goals. Axiomatic design minimizes the effective complexity of the engineered system while maintaining the system's effectiveness. Keep It Simple, Stupid—but not too simple.

Determining the effective complexity of a physical system obviously involves a judgment about what constitutes a regularity and what does not. That is, you must establish criteria that indicate when a bit is an "important" bit, a bit of regularity, and when it is "unimportant," a bit of randomness. In an engineered system, the important bits are those that have to take on particular values or else the system will not do what it's supposed to do. In evolved systems, such as a bacterium, it is less obvious which bits are important and which are unimportant. A simple criterion for figuring out whether a bit is important and so contributes to the effective complexity is to flip it and see what happens. If flipping the bit has a significant effect, it is important; if flipping the bit has no significant effect, it is unimportant. If the bit affects the bacterium's ability to survive and reproduce, then that bit contributes to the effective complexity of the bacterium. A bacterium's important bits are those that affect the bacterium's future in a significant way; the effective complexity of any system that exhibits purposeful behavior can be similarly measured. Any bit that affects the ability of the system to attain its purpose contributes to the system's effective complexity.

Of course, the definition of purposeful behavior is to some degree subjective. But suppose we focus on behavior that allows a system to (a) get energy and (b) use that energy to construct copies of itself. Living systems devote most of their time and effort to eating and reproducing. However one defines life, any system that can accomplish those two actions has gone a long way on the road to being alive. Once we identify as purposeful those behaviors that enhance the system's ability to get energy and use it to reproduce, then we can measure the effective complexity of all living systems and all systems that may someday be alive. As we'll see, effectively complex systems that get energy and reproduce arise naturally out of the underlying computational processes of the universe.

Why Is the Universe Complex?

Now that we have formally defined complexity, we can demonstrate that the universe necessarily generates it. The laws of physics are computationally universal, enabling the universe to contain logically deep systems and systems with high effective complexity. But we can also show that the universe *must* contain such complex systems. Let's look at the first information-processing revolution—the creation of the universe—in detail.

In measuring the complexity of the universe, we'll work within the current standard cosmological model. In this model, there is not enough matter in the universe to eventually slow and then reverse its expansion, causing it to end in a Big Crunch. It will continue to expand forever. Such a universe is spatially infinite, even at the very beginning. But we are interested in the computation that the universe performs—that is, the causally connected part of the universe, the part within the horizon, made up of bits that can talk to one another. Unless we are explicitly talking about what is happening beyond the horizon, we will adopt the usual practice and refer to the part of the universe within the horizon as "the universe."

The first information-processing revolution begins with the beginning of the universe. Before the beginning there is nothing—no space, no time, no energy, no bits. At the very instant of beginning, nothing yet has happened. The monkeys have not yet begun to type.

Observational evidence suggests that in the beginning the universe was simple. As far as we can tell, there may have been only one possible initial state, and that state was everywhere the same. If there were only one possible initial state at time zero, the universe contained zero bits of information. Its logical depth, thermodynamic depth, and effective complexity were also zero.

Now the universe begins to compute. One Planck time later (10^{-44} seconds), the universe contains one bit within the horizon. The amount of computation that can be performed on this bit in one Planck time is one op; that is, the universe's effective complexity and thermodynamic depth can be no greater than one bit, and its logical depth can be no greater than one op. The monkeys have typed one bit.

As the universe expands, the number of bits within the horizon grows and the number of ops accumulates. The maximum logical depth is limited by the number of ops, and the effective complexity and thermodynamic depth are limited by the number of bits. Though the universe is growing in complexity, it is still relatively simple. But the monkeys are typing away.

What is the universe computing during these early times? As usual, it is computing its own behavior. The universe computes itself. If we knew more about quantum gravity, we could reproduce the first few steps of the universe's computation on existing, man-made quantum computers, simple though they are. In fact, the computational theory of quantum gravity advocated earlier gives a straightforward picture of what the universe is computing. In this picture, the universe embarks on all possible computations at once.

Recall that quantum computers have the ability to perform many computations simultaneously, using quantum parallelism. Almost all input quantum bits are superpositions of 0 and 1. There is only one state that is 0 and one state that is 1, but there are an infinite number of possible input states that are superpositions of 0 and 1. Consequently, almost all one-qubit inputs to the quantum computer tell it to do this and that simultaneously.

Similarly, almost all two-qubit input states are superpositions of 00, 01, 10, and 11. If each of these four inputs instructs the computer to perform a specific computation, then almost all two-qubit input states instruct the quantum computer to perform these four computations in quantum parallel. And so it goes. As the number of input qubits grows, the universal quantum computer continues to embark upon all possible computations at once.

Even though the very early universe is simple, neither effectively complex nor logically deep, it has a glorious future ahead. The early universe is what Charles Bennett calls an "ambitious" system: even if it is not initially complex, it is intrinsically able to generate large amounts of complexity over time.

In the early universe, our quantum monkeys are typing in superpositions of all possible inputs. The computational universe interprets these inputs as instructions to perform all possible computations in quantum parallel. (This superposition of all possible structures is sometimes called the multiverse.) In one of these parallel quantum computations, it generates the particular complexity we see around us. As always, with monkeys typing into computers, structures that can be generated from short programs are more likely than structures that require long programs to produce.

The universe is computing. Bits are flipping. What are these bits? Bits in the early universe represent local values of energy density. For example, a 0 can represent a lower-than-average energy density and a 1 a higher-than-average energy density. Because of the simple, homogeneous nature of the initial state, the average density of energy is everywhere the same, but there are quantum fluctuations about that average. The quantum bits of the universe are in a superposition of lower density plus higher density. In terms of energy, the natural dynamics of the universe create regions in which the energy density takes on a superposition of different values.

As soon as the universe begins, its qubits begin to flip and interact. That is, as soon as the monkeys have begun to type their program by creating a quantum superposition, the laws of physics start to interpret that program. Recall that information, once created, tends to spread. Information is infectious. Because of the sensitivity of quantum bits to interactions with the other quantum bits in their surroundings, quantum information is particularly infectious. As noted earlier, this spreading of quantum information leads to decoherence, the severing of histories.

Take one qubit in a superposition of 0 and 1. This qubit registers 0 and 1 at the same time, by the ordinary laws of quantum mechanics. Now let this qubit interact with another qubit in the state 0—for example, by undergoing a controlled-NOT operation on the second qubit with the first qubit as control. The two qubits taken together are now in a superposition of 00 and 11: the quantum information in the first qubit has infected the second qubit. As a result of the interaction, however, the first qubit taken on its own behaves as if it registered either 0 or 1 but not both; that is, the interaction has decohered the first qubit.

As more and more interactions between qubits take place, quantum information that is initially localized in individual qubits spreads out among many qubits. As this epidemic of shared quantum information grows, the qubits decohere. As they decohere (one history no longer having an effect on the other), we can say that a particular region has either a higher energy density or a lower energy density. In the language of decoherent histories, we can start to talk about the energy density of the universe at the dinner table.

The next step in the computational universe is a crucial one. Recall that gravity responds to the presence of energy. Where the energy density is higher, the fabric of spacetime begins to curve a little more. As the fluctuations in energy density decohere, gravity responds to the fluctuations in the energy of the quantum bits by *clumping matter together* in the "1" component of the superposition.

In the computational-universe model of quantum gravity, the clumping occurs in a natural fashion: the contents of the underlying quantum computation determine the structure of spacetime, including its curvature. So a component of the superposition with a 1 automatically induces higher curvature than a component with a 0. When the qubit decoheres, so that it now reads either 0 or 1 but not both, the curvature of spacetime is either higher (in the 1 component) or lower (in the 0 component), but not both. In the computational universe, when qubits decohere and begin to behave more classically, gravity also begins to behave classically.

This mechanism for decoherence stands in contrast to other theories of quantum gravity, in which the gravitational interaction itself decoheres the qubits. No matter which theory of quantum gravity you adopt, however, the picture of the universe at this early stage is basically the same. Bits are being created and beginning to flip. Gravity responds to those bit flips by clumping matter about the 1's. Quantum bits are decohering, and random sequences of 0's and 1's are being injected into the universe. The computation is off and running.

In addition to making the earth on which we walk, gravitational clumping supplies the raw material necessary for generating complexity. As matter clumps together, the energy that matter contains becomes available for use; the calories we consume to stay alive owe their origin to the gravitational clumping that formed the sun and made it shine. Gravitational clumping in the very early universe is responsible for the large-scale structure of galaxies and clusters of galaxies.

This initial revolution in information processing was followed by a sequence of further revolutions: life, sexual reproduction, brains, language, numbers, writing, printing, computing, and whatever comes next. Each successive information-processing revolution arises from the computational machinery of the previous revolution. In terms of complexity, each successive revolution inherits virtually all of the logical and thermodynamic depth of the previous revolution. For example, since sexual reproduction is based on life, it is at least as deep as life. Depth accumulates.

Effective complexity, by contrast, need not accumulate: the offspring need not be more effectively complex than the parent. In the design process, repeated redesign to hone away unnecessary features can lead to designs that are less effectively complex but more efficient than their predecessors. In addition to being refined away, effective complexity can also just disappear. The effective complexity of an organism is at least as great as the information content of its genes. When species go extinct, their effective complexity is lost.

Still, life on Earth seems to have started from a low level of effective complexity, then exploded to produce the hugely diverse and effectively complex world we see around us. The computational capacity of the universe means that logically and thermodynamically deep things necessarily evolve spontaneously. Does the computational universe spontaneously give rise to ever increasing effective complexity? Looking around, we see vast quantities of effective complexity. But does the total effective complexity necessarily increase? Or might it at some point collapse? The effective complexity of human society seems quite capable of collapsing, for example, in the event of all-out nuclear war. When the sun burns out billions of years hence, life on Earth will be over.

How, why, and when effective complexity increases are open questions in the science of complexity. We can get a sense of the answers to these questions by looking at the mechanisms that generate effective complexity. We have defined purposeful behavior as that which allows systems to (a) get energy and (b) reproduce. The effective complexity of a living system can be defined as the number of bits of information that affect the system's ability to consume energy and reproduce. If we add to these two behaviors a third, to reproduce *with variation,* then we can look at the way in which effective complexity changes over time.

Any system, such as sexual reproduction, that consumes energy and reproduces with variation can both generate additional effective complexity and lose existing effective complexity. Of the varying copies constructed during reproduction, some will be better at consuming and reproducing than others, and those variants will come to dominate the population. Some variants will have greater effective complexity than the original system and some will have less. To the extent that greater effective complexity enhances the ability to reproduce, effective complexity will tend to grow; by contrast, if some variant can reproduce better with less effective complexity, then effective complexity can also decrease. In a diverse environment with many reproducing variants, we expect effective complexity to grow in some populations and decrease in others.

All living systems consume energy and reproduce with variation, but such reproducing systems need not be alive. At the very beginning of the universe, the cosmological process called inflation produced new space and new free energy at a great rate. Each volume of space spawned new volumes by doubling in size every tiny fraction of a second. Space itself reproduced. Variation was supplied by quantum fluctuations (those monkeys): As space reproduced, each offspring volume was slightly different from its parent volume. As matter started to clump together around regions of greater density, those regions accumulated more free energy at the expense of other, less dense regions. Billions of years later, the Earth formed in one of those regions of higher density. And billions of years after that, some piece of Earth evolved into us.

Life Begins

Biologists know a huge amount about how living systems work; ironically, they know less about how life began than cosmologists know about the beginning of the universe. The date of the Big Bang and its location (everywhere) are known to a higher degree of precision than the date and location, let alone the procedural details, of the origin of life. What is known is that life first appeared on Earth almost 4 billion years ago. It might have originated here or it might have originated elsewhere and been transported here.

Wherever life began, how did it begin? The answer to this question is the subject of hot debate. Here is one scenario.

We have seen that the laws of physics allow computation at the scale of atoms, electrons, photons, and other elementary particles. Because of this computational universality, systems at larger scales are also computationally universal. You, I, and our computers are also capable of the same basic computation. Computation can take place at the scale just above the atomic scale, as well. Atoms can combine to form molecules. Chemistry is the science that describes how atoms combine, recombine, and disassociate. Simple chemical systems are also capable of computation.

How does chemistry compute? Imagine a container, such as a small pore in a rock, filled with various chemicals. At the beginning of our chemical computation, some of the chemicals have high concentrations. You can think of these chemicals as bits that read 1. Others have low concentrations: these read 0. Just where the boundary between high and low concentration lies is not particularly important for our purposes.

These chemicals react with one another. Some that started out in a high concentration are depleted; the bits corresponding to these chemicals go from 1 to 0. Some that started out in a low concentration go to a high concentration; these bits go from 0 to 1. As the chemical reactions proceed, some bits flip while others remain the same.

This sounds promising. After all, a computation is just bits flipping in a systematic fashion. In order to show that a chemical reaction can perform a universal computation, all we have to do is show that it can perform AND, NOT, and COPY operations.

Let's start with COPY. Suppose that chemical A enhances the production of chemical B, so that without lots of A around, the level of B remains low. If there is a low concentration of A and a low concentration of B, then the concentrations of A and B remain low. If the bit corresponding to A is 0 initially, as is the bit corresponding to B, then these bits remain 0. That is, $00 \rightarrow 00$. Similarly, if there is a high concentration of A and a low concentration of B to begin with, then the chemical reaction gives rise to a high concentration of A together with a high concentration of B. That is, if the bit corresponding to A is 1 initially, and the bit corresponding to B is 0, then these bits both end up 1. $10 \rightarrow 11$. The reaction has performed a COPY operation. The bit corresponding to A is what it was before the reaction and the bit corresponding to B is now a copy of the bit corresponding to A. Note that in this process, A has an effect on whether or not B is produced, but A itself is not consumed in the reaction; in chemical terms, A is called a catalyst for the production of B.

NOT is produced in a similar fashion. Suppose that instead of enhancing the production of B, the presence of A inhibits the production of B. In this case, the reaction leads to B's bit being the opposite of A's bit; that is, B's bit is the logical NOT of A's bit.

How about AND? Suppose chemical C goes from a low concentration to a high concentration if and only if there are high concentrations of A and B around. Then a reaction that starts out with C in low concentration (its bit is initially 0) leads to a high concentration of C if and only if both A and B are in a high concentration (i.e., if and only if A's bit and B's bit are both 1). After the reaction, C's bit is the logical AND of A's and B's bits.

Chemical reactions can readily produce AND, NOT, and COPY operations. By adding more chemicals to the set, such logic operations combine to produce a set of reactions corresponding to any desired logic circuit. Thus, chemical reactions are computationally universal.

In general, as the chemicals in the pore in the rock react, some are catalysts for the initial set of reactions and some of the products of these initial reactions are catalysts for yet further reactions. Such a process is called an "autocatalytic set of reactions": each reaction produces catalysts for other reactions within the set. Autocatalytic sets of reactions are powerful systems. In addition to computing, they can produce a wide variety of chemical outputs. In effect, an autocatalytic set of reactions is like a tiny, computer-controlled factory for producing chemicals. Some of these chemicals are the constituents of life.

Did life begin as an autocatalytic set? Maybe so. We won't know for sure until we identify the circuit diagram and the program for the autocatalytic set that first started producing cells and genes. The computational universality of autocatalytic sets tells us that some such program exists, but it doesn't tell us that such a program is simple or easy to find.

Many Worlds, Again

In his 1997 book *The Fabric of Reality*, the physicist David Deutsch eloquently defends the Many Worlds theory of quantum mechanics in terms of quantum computation. Before wrapping up, let's briefly look at the senses in which other worlds, à la Deutsch and Borges, can exist.

The universe we see around us corresponds to just one of a set of decoherent histories; that is, what we actually see when we look out the window is just one component of the superposition of states that make up the overall quantum state of the universe. The other components of this state correspond to "other worlds," worlds in which the tosses of the quantum dice turned out differently. The set of all possible worlds together constitutes the multiverse. I'll leave it up to the reader to decide whether or not these other worlds exist in the same way as ours does. Whether they exist or not, as long as they are decoherent, these worlds can have no effect on our own.

Note that our history is effectively complex. Like other histories in the decoherent set, ours is the result of many, many throws of the quantum dice. (About 10^{92}, to be exact.) Nonetheless, the overall quantum state of the universe remains simple: the universe begins in a simple state and evolves according to simple laws.

How can our history, which is just part of the entire state of the universe, be more effectively complex than the whole? There is nothing particularly paradoxical about it: the set of all billion-bit numbers is simple to describe, but almost any given number in the set requires a billion bits to describe. The same principle holds for the state of the multiverse. A given component of the superposition can require about 10^{92} bits to describe, while the state as a whole requires only a few. In the case of the computational universe, the overall state is simple: the multiverse is performing all possible computations in quantum parallel. But to specify any single one of those computations requires summoning the bits that correspond to the program for that computation. A given computation may take many bits to specify.

As the multiverse computes, every possible computation is represented in quantum parallel in its overall state. The probability of any given computation is equal to the probability that monkeys type out its program. According to the Church-Turing hypothesis, every possible mathematical structure is represented in some component of the superposition. One such mathematical structure is the structure we see around us, every detail that we observe, including the laws of physics, chemistry, and biology. In other components of the superposition, the details are different. In some component, everything else is the same, but I have brown eyes instead of blue. In some component, it may even be that some features of the Standard Model for elementary particles, such as the masses of quarks, are different from those in other components of the superposition.

There is a second way that all possible mathematical structures can be generated. Current observational evidence suggests that the universe is spatially infinite: it extends forever outside of the horizon. If this is the case, then somewhere, sometime, it will generate every possible mathematical structure. These structures can come to exist within our branch of the superposition; at some point in the future, they will come within our horizon and can affect us. Somewhere out there, there are exact copies of you and me. Somewhere else, the copies exist but are imperfect: I have brown eyes instead of blue. At some point in the future, information about these distant copies will enter our horizon. But the stars will have gone out long before. As Boltzmann might have said, if you're interested in communicating with such other worlds, don't hold your breath.

By contrast, if you're interested in communicating with life on other planets, then you might be in luck. For the same reason that we know that the laws of physics support computation (we have computers), we know that they support life (we are alive). But we don't know the probability of life arising spontaneously on some other planet, nor do we know the probability that life, once established on one planet, could be transported to another. The chances of communicating with living creatures from another planet depend crucially on these odds. Someday we may know enough about how life arose to calculate them; until then, you have to ask yourself, "Do I feel lucky?"

The Future

How long can computation continue in the universe? Current observational evidence suggests that the universe will expand forever. As it expands, the number of ops performed and the number of bits available within the horizon will continue to grow. Entropy will also increase, but—because as the universe gets bigger it takes longer and longer to reach thermal equilibrium—actual entropy will increase at a slower rate than maximum possible entropy. As a result, the number of calories of free energy available for consumption within the horizon will increase.

So far, the news seems good. The problem is that while the total amount of free energy within the horizon continues to grow, the density of free energy—the amount of free energy per cubic meter—is decreasing. That is, there are more calories out there, but they're getting harder and harder to collect. Trillions of years from now, the stars will have burned through their store of nuclear fuel. At that point, our descendants, should they still be around, could harvest energy by collecting matter and converting it into usable energy, a strategy analyzed in detail by Steven Frautschi of Caltech.* The maximum amount of free energy

*"Entropy in an Expanding Universe," *Science* 217, no. 4560 (Aug. 13, 1982): 593–99.

that could be extracted is $E = mc^2$, where m is the mass of the collected matter. (Of course, some fraction of the energy will be lost due to inefficiency of extraction.)

By scavenging farther and farther afield, our descendants will collect more and more matter and extract its energy. Some fraction of this energy will inevitably be wasted or lost in transmission. Some cosmological models allow the continued collection of energy ad infinitum, but others do not.*

A more parsimonious strategy for eternal life is to make do with a finite amount of energy, as proposed by Freeman Dyson, of the Institute for Advanced Study.† After all, the total number of ops that can be performed is proportional to the amount of energy available times the amount of time for which it is available. If time goes on forever, a finite amount of energy should suffice to compute forever. Unfortunately, whenever an op is performed, some of the energy will be wasted due to errors and inefficiency. Eventually the supply of energy will dwindle and approach zero. Dyson points out that despite the dwindling of stores of energy, life could still go on as long as it is willing to slow down.

Suppose that each time this future life-form performs an op, all the energy that was used to perform it is dissipated. This is the worst-case scenario. So the next time the being performs an op, there is less energy available. That's OK—the next op is just performed more slowly, using a smaller amount of energy. The available energy is gradually decreasing, but at a slower and slower rate. Similarly, the time taken to perform each

---

*For a pessimistic view of the ultimate future of life, see "The Fate of Life in the Universe," by Lawrence Krauss and Glen Starkman, *Scientific American* 281 (November 1999). The authors cite recent observations indicating that the expansion of the universe is accelerating. If this acceleration continues at the observed rate, eventually the amount of available energy within the horizon will go to zero. For a more optimistic view, see "The Ultimate Fate of Life in an Accelerating Universe," by Katherine Freese and William H. Kinney (http://arXiv.org/astro-ph/0205279). These authors anticipate that the expansion rate will slow and the amount of available energy within the horizon continue to increase.

†"Time Without End: Physics and Biology in an Open Universe," *Reviews of Modern Physics* 51, no. 3 (July 1979): 447–460.

op gets longer and longer. But as long as you keep on performing ops more and more slowly, you can still perform an infinite number of ops in an infinite time, using a finite amount of energy.

How about memory space? As the amount of available energy dwindles, the total memory space available in a given volume also drops. So to keep on increasing the amount of memory space available, our deathless life-form must spread its energy out over a larger and larger volume. In other words, if you want to live forever, you have to slow down and get fat (a strategy many people have already adopted).

The biggest potential problem with this slow-down/get-fat strategy is waste. You have to get rid of that used-up energy somehow. Fortunately, slowing down and getting fat also helps here: the slower you are, the less energy you have to dissipate, and the larger you are, the greater the surface area you have through which to dissipate it. You have to be careful, however, to expand slowly enough so that your average energy per bit (that is, your temperature) remains above the temperature of the surrounding universe. If the universe has an intrinsic minimum temperature, as suggested by some cosmological observations, then you're sunk. At some point, you'll just be swamped by the surrounding radiation. However, if the temperature of the universe keeps decreasing forever at a sufficiently rapid rate, as suggested by other cosmological observations, then you're cool: you can keep on processing information and increasing your memory space.

Supposing it can exist, what would such an ultimate life-form look like? It would expand to encompass first stars, then galaxies, then clusters of galaxies, and eventually, it would take billions of years to have a single thought. Attractive? It depends on your taste. But if you want to live forever, you have to expect to make a few sacrifices.

Being Human

We've looked at the hot past; we've looked at the dim and distant future. To conclude our discussion of complexity, let's return to the present.

Where do human beings fit in the computational universe? The innate information-processing capacity of the universe at the fundamental level gives rise to all possible forms of information processing. After the Big Bang, as different pieces of the universe tried out all possible ways of processing information, sooner or later, seeded by a quantum accident, some piece of the universe managed to find an algorithm to reproduce itself. That accident led to life. Life evolved by processing genetic information to try out new strategies for survival and reproduction. After trying out billions of strategies, some living systems eventually discovered sex, a technique that vastly increases the rate at which new evolutionary strategies and algorithms can be explored, because it speeds up the rate of genetic information processing. After billions of years of sex, living creatures had evolved all sorts of methods for getting and processing information—eyes, ears, and brains, to name a few.

Somewhere in the last 100,000 years or so, human beings hit upon language. Human language must have seemed an odd-sounding innovation to the other animals around. But by allowing the expression of arbitrarily complicated concepts, human language allowed people to process information in a highly distributed fashion. The distributed nature of human information processing in turn allowed people to cooperate in new ways, forming groups, associations, societies, companies, and so on. Some of these new forms of cooperation proved strikingly effective, as various forms of distributed information processing, such as democracy, communism, capitalism, religion, and science, took on a life of their own, propagating themselves and evolving over time.

It is the richness and complexity of our shared information processing that has brought us this far. The invention of human language, coupled with diverse social development, was a true information-processing revolution that has substantially changed the face of the Earth. It has been argued that the human brain sets us apart from other animals. We human beings are very attached to our brains. Without them, we would have no thought or perception (the same holds true for other animals with brains). Language allows us to connect the workings of our brains to the workings of other people's brains. Communication allows us to collaborate and compete in ever more complex ways. To paraphrase John Donne, no one is an island. Every human being on Earth is part of a shared computation.

It is the joint computation shared by all of human society that makes us special—if indeed we are special, that is. Human beings are not the only physical systems to participate in a complex, rich computation. As I've tried to show, every atom, every elementary particle participates in the huge computation that is the universe. At bottom, each bit of the universe is just a bit. In their ability to register and process information, all bits are equal.

Science has an uncomfortable way of pushing human beings from center stage. In our prescientific stories, humans began as the focal point of Nature, living on an Earth that was the center of the universe. As the origins of the Earth and of mankind were investigated more carefully, it became clear that Nature had other interests beyond people, and the Earth was less central than previously hoped. Humankind is just one branch of the great family of life, and the Earth is a smallish planet orbiting an unexceptional sun quite far out on one arm of a run-of-the-mill spiral galaxy.

We are, nonetheless, unique (as are bacteria; as are elm trees). What makes us unique is information—the bits of DNA that join us to monkeys, and the habits of language and thought that separate us from them. There is no separate substance, no *vis vitae* or vital force, that makes us living, breathing human beings. We are made of atoms, like everything else. It is the way that those atoms process information and compute in concert that makes us what we are. We are clay, but we are *computational* clay.

Universal Thoughts

Now that we are aware of the computational nature of the universe as a whole, it is tempting to ascribe to it a kind of cosmic intelligence, like

Laplace's divine "demon." There is nothing wrong with thinking of the universe itself as some kind of gigantic intelligent organism, any more than it is wrong to think of the Earth itself as a single living being (an idea known as the "Gaia hypothesis"). Note, however, that if you assert the intelligence of the universe, you cannot deny the brilliance of one of its greatest "ideas"—natural selection.* For billions of years, the universe has painstakingly designed new structures by a slow process of trial and error. Each "Aha" in this design process is a tiny quantum accident, whose consequences are elaborated by the laws of physics. Some accidents work out, others don't. After billions of years, the result is us, and everything else.

In the final analysis, to say that the world is alive, or that the universe thinks, is only a metaphor. After all, what are these thoughts of the universe? Some of the information processing the universe performs is indeed thought—human thought. Some of that information processing, like digital computation, can resemble thought. But the vast majority of the information processing in the universe lies in the collision of atoms, in the slight motions of matter and light.

Compared with what is normally called thought, such universal "thoughts" are humble: they consist of elementary particles just minding their own business. But humility is not the same as weakness. Quantum chaos can amplify slight motions until they become a hurricane. The microscopic dance of matter and light had the power to produce not just human beings, but every being. The collision of two atoms can—and does—change the future of the universe.

*To deny the evidence for natural selection is to insult the intelligence of the universe.

Personal Note: The Consolation of Information

My road to the concept of effective complexity has been a long and complex one. It began with my work with Heinz Pagels at Rockefeller, and it continued as my Ph.D. moved into view and I was offered a postdoctoral position with Rolf Landauer at IBM. Landauer was one of the founders of the field of physics of information. His motto, "Information is physical," is an underlying principle of this book: all information that exists is registered by physical systems, and all physical systems register information.

I was somewhat surprised by the job offer. I had gone up to IBM's Watson Laboratories in Yorktown Heights that fall to deliver a talk on Maxwell's demon. As part of my Ph.D. thesis ("Black Holes, Demons, and the Loss of Coherence: How Complex Systems Get Information and What They Do with It"), I had made a quantum-mechanical model of how one quantum system gets information about another and showed how such apparent violations of the second law of thermodynamics do not cause any actual violations. My talk had not gone too well. Landauer had been crusty. Charles Bennett had just published a definitive article on Maxwell's demon that year, articulating far better than I had the trade-off between information and entropy. Perhaps worse, I inadvertently insulted Gregory Chaitin at lunch by making a joke about people who believe in the healing power of crystals, unaware that he kept a large crystal in his living room because it helped him concentrate.

Nevertheless, here was the impressive job offer, and I made preparations to go. Shortly after Landauer's call, the head of my laboratory at Rockefeller walked into my office. "Murray Gell-Mann wants to talk with you on the phone immediately," he declared. By now I was wary of professors walking into my office and making pronouncements. Why on earth would Gell-Mann want to talk with me? I had never met him (the fistfight in the convent hadn't happened yet), and

I had no idea why the world's best-known physicist would be interested in any of the peculiar stuff I was doing.

I picked up the phone. "Where is your application to Caltech?" Gell-Mann demanded. He was working on problems of complexity and on foundations of quantum mechanics and was having a hard time finding a postdoctoral fellow to work with him on these subjects. He'd been searching for months for someone who had done a Ph.D. in these fields and had finally come across me. He would piece together a postdoctoral position for me, if I was willing to accept. I had gone from no job to two job offers in the course of a week. It was a tough decision to make. Because I had missed the usual Caltech hiring process, the salary that Gell-Mann could offer me for the first year was half of what IBM was offering. On the other hand, I was excited by the prospect of going to Caltech. In the end, I decided to go west. It was the summer of 1988. I packed my belongings into my ten-year-old Datsun and started driving.

My first stop was Santa Fe, New Mexico, where I attended the first Santa Fe Institute Summer School and met Gell-Mann in person for the first time. We drove up to Los Alamos together and spent the afternoon debating the concept of complexity. Gell-Mann is a striking sight, with curly white hair and an electric smile. He is a remarkable person with whom to have a conversation, for the simple reason that he has about ten times the normal expected knowledge on virtually any subject you care to mention. He does not hesitate to let you know if your own speculations on the subject are mistaken. At one point in our three-hour discussion I hazarded an opinion about an aspect of quantum mechanics with which I was not sufficiently familiar. "No," said Gell-Mann, his voice getting louder. "No!" Putting his forehead down on the table where we were sitting, he began pounding the table with his fists. "No! No! No! No!! *No!!!*" Here, I thought, was someone I could work with.

After a month in Santa Fe, I drove up to the Aspen Center for Physics to visit Heinz, who had a house outside Aspen, where he and Elaine and their two small children were spending the summer. Our work on thermodynamic depth had inspired debate in the scientific community and attention from science journalists. On a series of hikes in the Elk Mountains, we discussed our next work. Or rather, we spent a small fraction of the time discussing our next work; most of the time Heinz regaled me with wild stories about things other than physics. Atop Castle Peak, he spread his arms to the mountains around and declaimed, "I present to you all the wealth and all the beautiful women in the world."

Two days later, disaster struck. Heinz and I had decided to climb Pyramid Peak, a 14,000-foot pile of rotten rock in the West Maroon wilderness about ten miles from Aspen. The day was fine; we started early and, taking care to avoid falling rock, we were at the top by noon. On the way down, we had to make a brief traverse with a lot of exposure; if you fell, you fell a long way. We edged along a crack in the rock, Heinz in front. At the end of the crack, he hopped onto a saddle between two crags. Heinz's ankle had been weakened by polio when he was a child. When his foot hit the saddle, his ankle buckled. He fell and slid out of sight down an almost vertical gully.

I called. There was no response. Quickly, before I froze up, I jumped from the crack to the saddle. It was a small, slightly awkward hop. I called again and again to silence. The gully was too steep to descend on my own, and Heinz had had the rope in his backpack. I turned and ran down the trail to get help.

There is no happy ending. The sheriff took me to tell Elaine what had happened, while the mountain rescue crew went to look for Heinz in the hope that he might still be alive in a crevice. When they couldn't find him, they took me back up the mountain in a helicopter. We rose slowly up into the great bowl of Pyramid Peak, the central part of it a cliff of sheer and rotten rock rising half a mile to the summit. We found where Heinz had fallen. There was no crevice. He had fallen 100 feet, hit hard, and died—a quick, clean death. After impact, his body had tumbled 2,000 feet farther down the mountain, where we found him on a shelf of rock. Slowly, we sank down out of the bowl. Then I went to give Elaine the bad news.

Over the following months, I tried to make sense of what happened. My loss was nothing compared with Elaine's; still, between the shock and the sadness, I was nearly destroyed. After the funeral, I returned to Los Alamos, to work with Wojciech Zurek. I was living in a bed-and-breakfast overlooking a canyon. At night, Heinz's voice would course through my head and wake me from sleep. I'd jump out of bed, thinking he was in the room. Guilty for having survived, I went on a long hiking trip by myself in the Pecos Wilderness. I lost myself in the woods; for several days of hiking I had no idea where I was. I did solo climbs I should not have done. I sought out cliffs and peered down them, terrified, from the top.

I looked for some kind of comfort in work, but physical law, while absorbing, is short on comfort. All chance occurrences have their roots in quantum mechanics. The Many Worlds interpretation of quantum mechanics holds that for every accident that happens, there are a lot of other worlds in which it doesn't. Heinz's death was a tragic accident. He was an experienced climber and could have jumped from the crack to the saddle a thousand times safely. Just this time, he landed at slightly the wrong angle, and his ankle failed him. Here, in our world, he fell. Other worlds gave me no comfort. Like the protagonist of Kenzaburo

Personal Note: The Consolation of Information

Oe's novel *A Personal Matter,* I learned that "you can't make the absoluteness of death relative, no matter what psychological tricks you use."

But this world still affords a measure of consolation. In talking with Elaine, and with Heinz's friends John Brockman and Sharon and David Olds, I have learned more about Heinz and his life. In working on ideas that stemmed from our brief collaboration, including some of the ideas in this book, I have gained satisfaction in imagining Heinz's responses and criticisms. Consolation has gradually come from information—from bits both real and imagined. Heinz's body and brain are gone. The information his cells processed is wrapped up in the Earth's slow processes. He has lost consciousness, thought, and action. But we have not entirely lost him. While he lived, Heinz programmed his own piece of the universe. The resulting computation unfolds in us and around us: the vivid thoughts and outrageous behavior he impressed on us still flourish in our thoughts and behavior and have their own vivid and outrageous consequences. Heinz's piece of the universal computation goes on.

Acknowledgments

I would like to thank all my teachers, particularly my family and friends.

My wife, Eve, and daughters, Emma and Zoe, exhibited great tolerance during the writing of this book. My parents, Robert and Susan Lloyd, were the book's first readers and editors. My brothers Ben and Tom both contributed valuable questions, as did my nieces, nephews, cousins, uncles, aunts, and in-laws.

My friends at MIT and elsewhere gave many useful comments and criticisms, in particular Charlie Bennett, Paul Davies, and the members of the Moses seminar: Joel Moses, Bob Berwick, Robert Fano, Gadi Geiger, Jay Keyser, Tom Knight, Sanjoy Mitter, Arthur Steinberg, and Gerry Sussman. Terry Orlando's graduate reading group patiently read the manuscript and told me what they thought; I listened. JR Lucas, Janet Brown, and Aram Harrow helped me track down elusive typing monkeys. Murray Gell-Mann taught me quantum mechanics and complexity, and Doyne Farmer made me bicycle up and down high mountains while discussing the relation between the two topics. Shen Tsai helped me with Mencius.

All my colleagues in the field of quantum information and computation contributed greatly to this book via their own scientific work. Much of the research on which this book is based was funded by the Cambridge-MIT Initiative, the National Science Foundation, the Army Research Office, the Defense Advanced Research Projects Agency, the Advanced Research and Development Activity, the Naval Research Office, and the Air Force Office for Scientific Research.

Marty Asher at Knopf was a patient and sensitive editor. Any style in the prose is due to Sara Lippincott, who turned my jotting into writing. John Brockman forced me to write the book in the first place, and Katinka Matson held my hand while I did it.

Finally, I'd like to thank those who are no longer here to be thanked, notably Heinz Pagels, Rolf Landauer, and Alexis Belash.

Further Reading

Discussions of the universe as a computer abound. In addition to Asimov's "The Last Question" (1956), see, e.g., H. R. Pagels, *The Cosmic Code* (Simon & Schuster, 1982), J. D. Barrow, *Theories of Everything* (Clarendon Press, 1991), and F. J. Tipler, *The Physics of Immortality* (Doubleday, 1994).

The idea that the universe might be a classical digital computer was put forth in the 1960s by Konrad Zuse and Ed Fredkin. Zuse's book is *Rechnender Raum* (*Schriften zur Datenverarbeitung*, Band 1, Friedrich Vieweg & Sohn, Braunschweig, 1969), translated as *Calculating Space* (MIT Technical Translation AZT-70-164-GEMIT, MIT [Proj. MAC], Cambridge, Mass. 02139, February 1970, http://www.idsia.ch/~juergen/zuse.html). Fredkin's work can be found at http://www.digitalphilosophy.org/. The particular type of computer that they proposed was a "cellular automaton." A cellular automaton consists of a regular array of cells, each of which contains one or more bits. Each cell updates itself from time step to time step as a function of its own state and that of its neighbors. The idea of universe as cellular automaton has more recently been popularized by Stephen Wolfram in *A New Kind of Science* (Wolfram Media, 2002).

For the mathematics behind monkeys typing on computers, see R. J. Solomonoff, "A Formal Theory of Inductive Inference," *Information and Control* 7 (1964), 1–22; G. J. Chaitin, *Algorithmic Information Theory* (Cambridge University Press, 1987); A. N. Kolmogorov, "Three Approaches to the Quantitative Definition of Information," *Problems of Information Transmission* 1 (1965), 1–11. Further discussion of the concept of algorithmic information and its relationship to the generation of complexity can be found the writings of Juergen Schmidhuber at http://www.idsia.ch/~juergen. See also Max Tegmark, "Is 'The Theory of Everything' Merely the Ultimate Ensemble Theory?" *Annals of Physics* 270 (1998), 1–51 (arXiv/gr-qc/9704009). For the relationship between algorithmic information and the second law of thermodynamics, see, e.g., W. H. Zurek, *Nature* 341 (1989), 119–24.

The idea that the undecidability problem and the halting problem are related to the problem of free will was suggested by Turing in his article "Computing Machinery and Intelligence," *Mind* (1950), 433–60. See also K. R. Popper, "Indeterminism in Quantum Physics and Classical Physics," *British Journal for Philosophy of Science* 1 (1951), 179–88. A classic paper on this topic is J. R. Lucas, "Minds, Machines, and Gödel," *Philosophy* 36 (1961), 112–27. A more recent exploration of free will is *Elbow Room: The Varieties of Free Will Worth Wanting,* by Daniel C. Dennett (MIT Press, 1984). A study of the implications of the computational ability of the universe for our ability to predict its behavior can be found in D. R. Wolpert, "Computational Capabilities of Physical Systems," *Physical Review E* 65, 016128 (2001) (arXiv/physics/0005058, physics/0005059).

A summary of the second law of thermodynamics and the nature of time asymmetry can be found in P. C. W. Davies, *The Physics of Time Asymmetry* (University of California Press, 1989). *Physical Origins of Time Asymmetry,* edited by J. J. Halliwell, J. Pérez Mercader, and W. H. Zurek (Cambridge University Press, 1996), is a collection of scientific articles on the subject. Many of the original papers on Maxwell's demon can be found in *Maxwell's Demon 2: Entropy, Classical and Quantum Information, Computing,* Harvey S. Leff, Andrew F. Rex (editors), Institute of Physics, 2003.

Many of the classic papers on quantum mechanics are collected, with commentary, in *Quantum Theory and Measurement* (ed. J. A. Wheeler and W. H. Zurek, Princeton University Press, 1983). A textbook on quantum mechanics with an emphasis on foundational issues is *Quantum Theory: Concepts and Methods* by A. Peres (Springer, 1995). The decoherent histories approach to quantum mechanics is described by Robert Griffiths in his book *Consistent Quantum Theory* (Cambridge, 2003). The way in which decoherence and chaos conspire to generate information is described in F. M. Cucchietti, D. A. R. Dalvit, J. P. Paz, W. H. Zurek, *Physical Review Letters* 91 (2003), p. 210403.

An introduction to quantum mechanics and quantum computation can be found in *A Shortcut Through Time: The Path to the Quantum Computer* by G. Johnson (Knopf, 2003). The standard textbook on quantum computers is *Quantum Computation and Quantum Information* by M. A. Nielsen and I. L. Chuang (Cambridge University Press, 2000).

Some of my work on the physical limits to computation and the computational capacity of the universe can be found in "Universe as Quantum Computer,"

Complexity 3(1) (1997), 32–35 (arXiv/quant-ph/9912088); "Ultimate Physical Limits to Computation," *Nature* 406 (2000), 1047–54 (arXiv/quant-ph/9908043); and "Computational Capacity of the Universe," *Physical Review Letters* 88, 237901 (2002) (arXiv/quant-ph/0110141). A popular account of quantum gravity is *Three Roads to Quantum Gravity* by L. Smolin (Perseus Books, 2002). A technical version of my theory of quantum gravity based on quantum computation is "The Computational Universe: Quantum Gravity from Quantum Computation," arXiv/quant-ph/0501135.

Accounts of the sciences of complexity can be found in *The Quark and the Jaguar: Adventures in the Simple and Complex* by Murray Gell-Mann (Freeman, 1995); *Emergence: From Chaos to Order* by John H. Holland (Perseus, 1999); and *At Home in the Universe: The Search for Laws of Self-Organization and Complexity* by Stuart Kauffman (Oxford, 1996). Charles Bennett's analysis of complexity and definition of logical depth can be found in "Dissipation, Information, Computational Complexity, and the Definition of Organization," in *Emerging Syntheses in Science*, edited by D. Pines (Addison Wesley, 1987), and "Logical Depth and Physical Complexity," in *The Universal Turing Machine: A Half-Century Survey* edited by R. Herken (Oxford, 1988), pp. 227–57. The complementary notion of thermodynamic depth is described in S. Lloyd and H. Pagels, "Complexity as Thermodynamic Depth," *Annals of Physics* 188 (1988), 186–213.

Index

Page numbers in *italics* refer to illustrations.

www.vintage-books.co.uk

23/8/07